A History of Italian Wine

Manuel Vaquero Piñeiro
Paolo Tedeschi • Luciano Maffi

A History of Italian Wine

Culture, Economics, and Environment in the Nineteenth through Twenty-First Centuries

palgrave
macmillan

Manuel Vaquero Piñeiro
Perugia, Italy

Paolo Tedeschi
Brescia, Italy

Luciano Maffi
Milano, Italy

ISBN 978-3-031-06096-0 ISBN 978-3-031-06097-7 (eBook)
https://doi.org/10.1007/978-3-031-06097-7

This Palgrave Macmillan imprint is published by the registered company Springer Nature Switzerland AG.
The registered company address is: Gewerbestrasse 11, 6330 Cham, Switzerland

Contents

List of Figures

List of Images

List of Tables

CHAPTER 1

Why a Comprehensive History of the Italian Wine?

In ancient times, the Italian peninsula was called Enotria, that is "the land of wine". Origins of this beverage are lost in the folds of the earliest urban civilisations, and this created a wide and relevant narrative in which facts, myths, and divinities mingle (Lukacs 2012; Unwin 1996). Even if the analysis only concerns the period from the birth of the Kingdom of Italy in 1861 to the new millennium, an economic history of wine in Italy represents a challenge, an effort to arrive at a synthesis from an abundance of material, research leads, and scenarios to reconstruct. Authors' goal is to illustrate the evolution of Italian viticulture and winemaking under the same regulations from the first laws concerning the Italian wine sector to the twenty-first century with the effects related to the great globalisation of wine markets, the strong technological innovations concerning oenology. The historical success of the wine depends on both religious factors and the use of this beverage for parties and, in general, for creating a very good ambiance: so, the wine always had an important place in all European countries and not only in the region where the vines were cultivated. In these latter, wine represented a relevant product for the development of the countryside. Few European rural products had the same economic relevance as the wine: for example, the citrus fruits or the leaves of the mulberries, but they progressively declined. The wine was important for the society and the economy of the Italian peninsula in the ancient times: it continues to guarantee relevant incomes to Italian winemakers and State, and it progressively enlarges its positive effects to other economic sectors

M. Vaquero Piñeiro et al., *A History of Italian Wine*,
https://doi.org/10.1007/978-3-031-06097-7_1

(from the industrial production of fungicides and new modern wine casks to the oenological tourism)—so, its economic history is surely interesting.

The book examines a period of 160 years in which the Italian wine sector was able to strongly modernise and renovate itself adapting its products to international trends and progressively building the current excellent reputation of the Italian wine in the world markets. In the second half of the nineteenth century, most of wines made in the Italian peninsula had a local market or they were used in foreign countries to "cut" the local wine increasing its alcoholic content (Maffi et al. 2021b). In the new millennium, Italy is one of the main world producers (the first one in the last years), and its high-quality wines are exported everywhere. This book exactly aims at analysing the stages, actors, tools, and stimuli changing the viticulture and wine production and retailing systems and allowing the Italian oenological sector to reach this extraordinary result.

In the first major agricultural survey conducted after the birth of the Kingdom of Italy, Senator Stefano Jacini illustrated different agricultures existing in Italy and strictly linked to the different climate, history, institutions, social conditions, and so on (Jacini 1884; Biagioli 2014). Despite the variety of contexts that made Italian agriculture an extremely diverse reality, there were (and continue to be) some elements matching the divergences and composing a homogeneous scenario concerning some relevant aspects. One of the main common points was (and continues to be) represented by the widespread presence of viticulture: the catalogue about the Kingdom of Italy drafted for the Paris exhibition of 1862 put in evidence the presence of vineyards in every part of the Italian peninsula, from the Alpine valleys to the plains up to the heart of the Mediterranean in the island of Pantelleria. This obviously implied (and continue to imply) a great variety of grapes and forms of cultivation (promiscuous or specialised, row or sapling) depending on the heterogeneity of the location of the vineyards (plains, hills, mountains, terraces overlooking the sea, "married" with mulberries, in arable land cultivated with cereals). This wide and heterogeneous panorama existed at the end of the nineteenth century when the arrival of the phylloxera, the start of the Italian industrialisation and the changes in wine consumptions imposed the progressive substitution of the winemaking practices existing in the Italian countryside.

The statements collected at the beginning of the 1870s by the committee responsible for carrying out the first Italian industrial survey agreed on the low quality of the wines consumed by most of the Italian people. The production was purely family-based or served to supply the nearest urban market: only richest people had the possibility of consuming good imported wines, mainly French, German, and Hungarian ones. Differences depended on prices and customs: 'indigenous' wines were made to meet the local consumers' taste, while foreign wines often met with 'little sympathy' because they were not 'strong enough'. Many reasons got wine producers less attentive to the quality of their products, and they ranged from fiscal to agronomic or technical ones. There existed a very negative view of the consumption duty, which was applied at the gates of each individual town and favoured sales concerning the poor "ordinary wines of the country" to the detriment of luxury wines arriving from abroad. The predominance of low-value ordinary wines did not stimulate producers to improve their wines. Finally, wines mixed with water were exempt from payment of duty, as they did not have the characteristics of real wines: this guaranteed a good market for low-quality wines to mix with water.

Today, after a long and difficult search for balance and the regulation of a sector strongly depending on the trend of harvests (Gaeta and Corsinovi 2014), the situation appears radically changed and the results achieved are surprising. Italy shares the world record in wine production with France: 369,000 farms cultivating wine grapes, 55,000 establishments winemaking and 8000 bottling (Corsi et al. 2019, 48) allowed to produce, in 2019, 51,239,000 hectolitres, that is 17 per cent of all wine produced in the world. So, in terms of volumes of wine produced, it is possible to underline a predominance that has lasted uninterruptedly since the end of the nineteenth century. However, the main changes concern the evaluation and position of Italian wine in the world markets: it in fact achieved a leader's role in terms of turnover and image. Italy also leads the world in terms of the number of wines having a designation of origin; there are 626, far outstripping its main competitors, France and Spain, having 381 and 90 respectively. The Italian wine sector is in third place for turnover in the Italian agri-food sector and, moreover, it represents the main exports for the Italian primary sector.

In Italy, there exist 380,000 wine companies representing 23 per cent of farms. The cultivated area of specialised vines is about 665,000 hectares, the 5 per cent of the Italian agricultural surface: 48 per cent is dedicated to DOC (Controlled Designation of Origin) and DOCG (Controlled and Guaranteed Designation of Origin) wines. Exports amount to about 20 million hectolitres, the 21 per cent of the world market with a value that exceeds 6 billion euros. Statistics show the relevance of the wine sector for the Italian economy and in particular put in evidence the great increase during the second half of the twentieth century when the wine producers experienced a significant transformation and modernisation which allowed to overcome former limits. This process permitted the passage from a backward wine industry to the modern one existing in the new millennium and characterised by the presence of excellent and distinctive oenological products helping the building of the Italian food and drink myths (Campbell and Guibert 2007; Teughels and Scholliers 2017; Charters 2006; Oddy and Drouard 2016).

Despite the important achievements of the Italian oenological sector and its relevance for the Italian economy, the wine is scarcely present in the national historiographic debate. Some important synthesis works are available (Corsi et al. 2018, 2019; Federico and Martinelli 2018), but there is a lack of studies on the historical evolution of all aspects contributing to the transformation and success of the Italian wine sector. This absence in fact emerges from the comparison with many researches dedicated to wine at international level or to single countries. Recent relevant books (Anderson 2004; Anderson and Pinilla 2018; Anderson et al. 2017) dedicated chapters to the analysis of the Italian case, demonstrating its international relevance. Other books highlighted many aspects of Italian viticulture and wine industry, but the analysis was essentially related to the French case (Loubère 1978) or was not detailed because the main topic was represented by the wine sector in the world (Simpson 2011). So, there are no books entirely concerning the economic history of Italian wine as it happens in the French and Spanish cases (Loubère 1990; Bodinier et al. 2014; Pan-Montojo 1994; Carmona, Colomé, and Pan-Montojo 2001; Paul 1996).

The Italian great territorial fragmentation of wine production in fact gets more difficult for analysis regarding the whole country: as studies about regional viticulture in Italy show, there are some regions producing wines having a high international prestige since the end of the nineteenth century (Tuscany, Piedmont, Sicily) while other terroirs became relevant in the last decades of the twentieth century (Fig. 1.1). However, in this context presenting a huge variety of cases and historical paths, the oenological sector of all regions had to respect the same laws, and it was pushed or hindered by the same public support or fiscal regulations. So, this book pursues the goal of illustrating the evolution of the Italian wine sector without forgetting that a real Italian wine does not exist: the book in fact indicates how a national framework was created that helped the different oenological regions to improve their products and make them more and more diverse and of high quality.

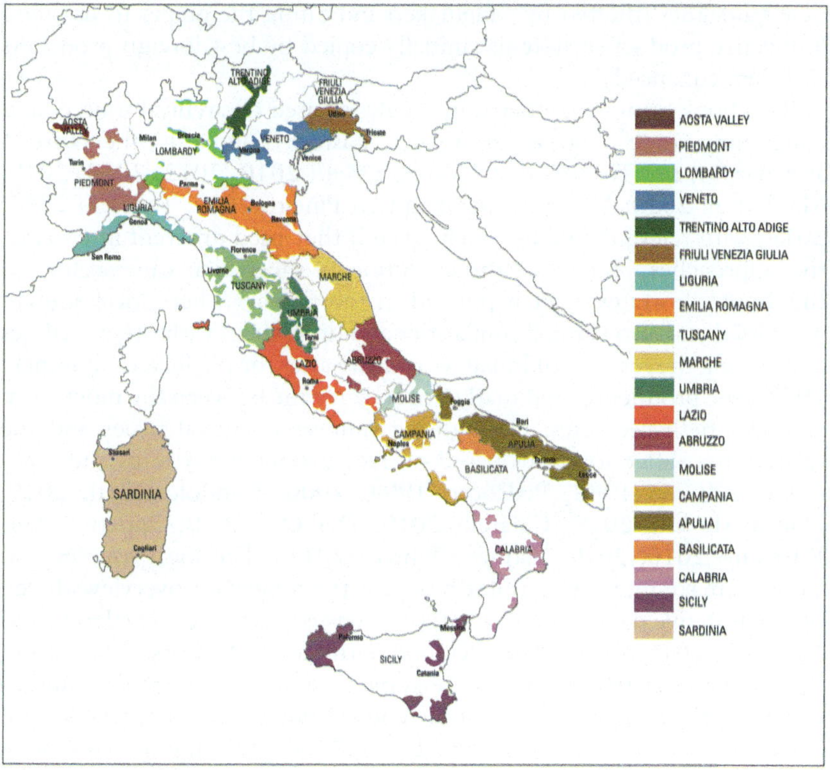

Fig. 1.1 Map of Italian winegrowing areas (2021). (Source: Our elaboration)

Until the 1960s, the relevance of the local tradition strongly influenced the production and the commerce of Italian wine. However, there existed some common productive systems and rules: the first depended on the network of information concerning oenology which was available since the end of the eighteenth century for the landowners who invested to improve the quality of their vines and wine; the second depended on the existing laws regulating the wine production and trade. It was the creation of the Common Agrarian Policy and the need to characterise and protect against fraudulent imitation of the Italian agricultural products that introduced and regulated the use of the trademark "Italian wine" to define some very different wines. These latter had (and continue to have) one common characteristic only, that is the research of a high quality based on new guidelines (drafted by winemakers indicating the grapes to use) and innovative productive systems (initially copied by best foreign producers and then enhanced).

This book brings together some analysis based on archival sources and concerning some oenological areas as Eastern Lombardy, the Oltrepò and Barolo, the Tuscany and Umbria (Maffi 2010, 2012, 2018, 2019; Maffi et al. 2020; Mocarelli and Vaquero Piñeiro 2019; Tedeschi 2017, 2019) with relevant studies which, even if they used different interpretative approaches, concerned the evolution of the Italian wine sector. So this book illustrates what happened in the terroirs where local sources were used and also verified similarities and differences with other analyses concerning the renewal of Italian viticulture after the phylloxera (Zaninelli 1977), the production and market of Italian wine between the nineteenth and twentieth centuries, highlighting numerous critical issues and the related innovative solutions based on new productive systems and technologies (Cova 1988; Pedrocco 1994, 2000; Dandolo 2010, 2019; Conca Messina 2019; Cafarelli 2019; Dell'Osa 2019; Rapetti 1984; Ritrovato 2010, 2019; Vaquero Piñeiro 2012). The local sources also allow to make interesting reflexions with the important overviews based on quantitative data related to national census and surveys (Federico and Martinelli 2018; Corsi, Pomarici, and Sardone 2018; Corsi, Mazzarino, and Pomarici 2019): it is in fact possible to note that public data underestimated the real production of the countryside because farmers did not declare correct data on their crops to avoid the increasing of taxes. So, a part of the actual wealth generated by the rural sector disappeared by official statistics and this explains, for example, why yields in Eastern Lombardy between the end of the nineteenth century and the twentieth one were "strangely" similar to the ordinary ones recorded at the half of the nineteenth century, while the technological evolution allowed to significantly improve harvest grapes and yields (Tedeschi 2008).

This book shows the main factors favouring the recourse to the new production system and the consequent improvement of the quality of the wine as well as the related enlargement of the market: from the production for self-consumption and for local markets (in particular those of the towns nearby the terroir) to the conquest of foreign ones. The formation of the modern Italian wine sector is the result of a slow and complex evolution concerning the number and the attitudes of producers, that is their investments and care for the final product. The book illustrates the disappearance of landowners and sharecroppers cultivating vines in small plots of land where there normally existed other productions (cereals between the row of vines or vineyards "married" to mulberries) and the arrival of investors renewing and selecting the vines, modifying the systems used for the grape harvest and for the conservation of the wort. During the twentieth century, new investments allowed to build new dry cellars wet, to increase the number of demijohns, to reduce the dimension of barrels (the *barriques* in wood oak giving more oxygen during the fermentation), and to use new giant tanks (in stainless steel or fibreglass). Other important changes concerned the quality of the wine: producers progressively substituted the mellow and heavy-bodied wine (which was realised in the nineteenth century and was mainly addressed to taverns supply) with some high-quality wine (which was and continues to be sold in bottles indicating the origins of the product) prepared to meet all consumers' taste and in particular the foreign ones.

Only the regional analysis allows to explain the factors originating these changes: a better knowledge of the local vineyard and the related wine production in fact allows to put in evidence and give more value to the human impact on the terroir, and by this way, to highlight the relevance of the oenological sector for the landscape and for the cultural heritage (linked to the winemaking and wine consumption). However, it is important to underline the State intervention in favour of the improvement of studies on ampelography and winemaking as well as public aids for the exports (Maffi et al. 2021a). The wine arriving on the market is in fact the result of the choices made by producers and public institutions (Colman 2008): so, the economic history of the Italian wine considers all actors who influenced the evolution of the productive and retailing systems and allowed Italian wines to progressively conquest the international markets. At the end of the nineteenth century, despite the availability of large quantities of vines and wines, the goal to improve the Italian wine quality and to export it to international markets was distant, uncertain, and full of obstacles. At that moment, the French example seemed unattainable. At the beginning of the new millennium, the situation was considerably changed: the success of Italian wine in all world markets was evident and the basic elements of this result were the increasing competition and the continuous renewal of the productive and retailing systems: these elements strongly depended (and continue to depend) on the integration of the public and private projects and on the related availability of financing for investments in innovation (Anderson and Pinilla 2021). Therefore, one of the key of Italian wine success was represented (and continue to be) by the long-standing public-private partnership.

This explains why the book dedicates a wide space to the analysis of both private entrepreneurs' attitudes and public administrations' policies. Moreover, it explains why authors analyse the Italian wine even if a real Italian wine has never existed because of the relevant regional differences existing in a country having different climate areas (from the Alpine valleys to the Morainique hills near the lakes, from the hills of the central Italy to the great isles in the Mediterranean sea). Italian public institutions, in fact, allowed winemakers to create and develop the numerous Italian wines sold around the world: for example, national laws and public investments strongly supported winemakers, regulated the competition in the wine sector and favoured the diffusion of the oenological education.

The book also puts in evidence the relevance of exogenous factors and the consequent reactions of Italian producers and public administrations: for example, in the second half of the nineteenth century some fungal diseases and insects (powdery mildew, peronospora, and phylloxera) seriously damaged vineyards and grapes and obliged producers to renovate their vines and cultivation systems. The related costs made a selection of the producers and vines favouring the improvement of the wine quality: even if few producers actually oriented their activity towards the high-quality production, the average quality of wine increased in all Italian regions. Another example is related to the transmission of the know-how which, until the twentieth century, depended on the existing networks related to the knowledge in agronomics: these networks were international but it is possible to evaluate their impact on local producers only if the analysis is regional. From the nineteenth century the creation of new agrarian schools and the Comizi Agrari (new public institutions created for improving the rural activities), the publication of agricultural journals, and the organisation of lessons and conferences allowed to diffuse all news concerning innovations in viticulture, production and conserving wine, and how to prevent (or to take care) of diseases. Some competitions and exhibitions for scholars and producers were also organised to improve the productivity of the vines and the resistance of the wine to the transport. The agrarian almanacs and calendars also contributed to the diffusion of the knowledge concerning the ampelography and winemaking: they were in fact diffused in all countryside. Studies about regional cases and in particular the use of local sources favoured the correct evaluation of actual transfer of the know-how which was realised and the real impact of the diffusion of the new knowledge in winemaking: besides, they help to understand why there were different reactions and, for instance, why, until the 1960s, the number of producers really improving their wine was limited and few investors were interested in the oenological sector. Regional analysis also allows to observe the latest trends resulting from greater sensitivity to biodiversity and the environment: climate changes are modifying the world geography of wine production, and they obviously influence Italian productions which are characterised by very different areas for the latitude. For example, even if they both produce passito wines, the isle of Pantelleria and Southern Sicily are 37–38 degrees North, while alpine valleys in Lombardy and Trentino are 46 degrees North: it is evident that they have to face different climate change effects concerning vines, insects and fungal diseases.

The book also investigates the role of international markets from the end of the nineteenth century to the new millennium. They were crucial in the modernisation of the Italian wine sector, in particular the American one where Italian migrants' demand stimulated the presence of the best Italian wines and also the improvement of the taste to enlarge the market to other consumers (Cinotto 2012, 2013; Hancock 2009; Law and Libecap 2006; Lapsley 1996; Heintz 1995). The conquest of foreign markets, questioning the first position of French wines, forced Italian producers to adopt new marketing techniques and transform their wine from a simple drink to a socially valued product. They started to analyse the characteristics of foreign market and progressively modernise the advertising, the industrial design, and the logistics for the wine exportation. The new division between people working in winemaking phase and people operating in the marketing phase favoured the diffusion of Italian wines and also allowed the creation of new professional skills and job places. This process was supported by the Italian public institutions: they in fact founded agencies managed by wine specialists (in New York, Buenos Aires, London, etc.) who stimulated the export of Italian wines.

The book in fact illustrates the role and impact of the legislation concerning the wine: the Italian State was interested in the alcoholic beverages not only for fiscal aspects (the taxes on consumption and tariffs for the importation) or for the control of the main negative effects of the excessive consume. The Italian State regulated the oenological sector because it guaranteed relevant exports taking profit from the existence of many wines, which could meet consumers' taste all over the world. As it arrived in Italy during the crisis linked to the phylloxera (at the end of the nineteenth century) or during the "economic miracle" (from the half of the 1950s to the early 1970s), the lobbies of the wine producers were able to orient the legislation in their favour limiting the public care for the consumers: however, this also happened because public institutions were conscious of the economic relevance for all Italian citizens of the economic results of the Italian wine sector and, moreover, consumers could change their taste and, by this way, to influence and reduce wine producers' power on the market. When the general improvement of the wages and earnings of the 1960s favoured a more elastic demand and oriented consumers' preferences towards medium and high-quality wines, producers in fact had to change their production and retailing. A similar situation arrived in the early 1900s when the increase in cost of winemaking (because of the need of grafting native vines with American ones) also obliged to improve the quality to justify new prices. Finally, the new European legislation concerning the indication of the origin of the product (coming from the French Appellation d'Origine Contrôlée) also had relevant effects: even if there always exist different markets depending on the price and the medium quality, new rules promoted the increasing of the European wine quality and, moreover, pushed consumers to associate the quality with the provenience of the product. This obliged wine producers to improve (or at least keep unchanged) the quality of their wine and to indicate its origin. It is exactly the new relevance for the origin of the product and its relation with the terroir which built consumer loyalty and then implemented the oenological tourism increasing the public institutions interest for the Italian wine sector and their help for the diffusion of the image of the "Italian wine", which became an icon of Italian Wine and Food even if in reality was (and continues to be) represented by very different wines. The Italian wine has never existed, but a national image and dimension of wine has been created and it has progressively included all high-quality wines having different targets, that is different consumers' tastes and budgets to spend (as it has happened for the Italian fashion).

The book also analyses the transformation of winegrowing and winemaking of Italian table wines characterised by a marked vocation for mass consumption and a widespread presence in the country. Besides, it gives new elements for the existing debate on the use of the concept of "quality" (and its progressive improvement) applied to wine (Garcie-Parpet 2009; Stanziani 2004). Furthermore, it increases the knowledge of the Italian rural society from the half of the nineteenth century to the new millennium: it in fact illustrates the trajectory of different production areas indicating their cultural and landscape values as well as main technical, social, and anthropological changes arriving in the Italian countryside.

So, this book also indicates some relevant effects of the changes concerning the Italian wine sector on the Italian society and economy. Following an evident economic history approach, it considers that the transformations experienced by the wine sector represent a relevant example of the overall modernisation of the Italian economy in the analysed period. In opposite, few attention is paid to special wines (as Marsala, vermouths, bitters) (Hancock 2009; Simpson 2011, 24–29), which, during the analysed period, always constituted very specific niches of market and experienced only a part of the transformations concerning the Italian table wines (Federico and Martinelli 2018). In fact, these cases are very interesting in particular for the business history: however, they do not represent what has happened to the Italian wine sector.

The main keys to the world success of Italian wine, a goal achieved despite premises that were by no means taken for granted, are indicated in four chapters. The first one establishes a link between the first jolts of oenological "renewal" which began at the end of the nineteenth century and the great turning point of the second half of the twentieth century. The Italian wine sector, abandoning old production and commercial logics, imposed itself on the world market for the quality and quantity it achieved. A new specific style of consumption was created and diffused in the world, filling the Italian wine, a beverage normally charged with cultural and social values, with new evocative and landscape values (Hall et al. 2004; Harvey et al. 2014).

The second chapter analyses the role of the Italian public institutions. From the 1870s to the First World War, the central, provincial, and municipal public administrations promoted the modernisation of Italian viticulture and winemaking, as well as the internationalisation of Italian wines. They played a fundamental role for: (a) the transmission and transfer of knowledge and information about the production system (winegrowing and winemaking); (b) the control of the final product to improve its quality and avoid alterations representing a danger for consumers and also for Italian producers' reputation; (c) the distribution of Italian wines on international markets, giving wine producers all the information about customers' tastes and the potential demand concerning Italian wines. During the analysed period, the initiatives of public institutions in favour of the creation and the diffusion of new national wines which were adapted to the international markets were numerous. They played an important role in creating a favourable environment for the modernisation of the sector and the improvement of competitive commercial strategies: they fostered the formation of modern oenological entrepreneurship and, at the same time, lead to the formation of oenological cooperatives. Furthermore, they helped the creation of a national oenology, which progressively increased the quality of Italian wines and their reputation in the world.

The third chapter presents the role of private entrepreneurship and cooperatives: it in fact analyses the formation and development of wine entrepreneurship in Italy between the 1880s and the beginning of the new millennium. Before the First World War, the international market made a decisive contribution to the modernisation of Italian wineries. The most relevant result was the formation of new entrepreneurs resulting by the fusion of noble families and the commercial bourgeoisie: the progressive elimination of rentiers in favour of new landowners investing to improve quantity and quality of their wine favoured the consolidation of dynamic entrepreneurial groups operating in the wine and spirits sector. Large entrepreneurial family groups progressively become stronger: their wineries merged to create businesses with financial and organisational means that allowed them to be competitive everywhere. Some of these wine companies, born during the last decades of the nineteenth century and open to global markets, demonstrated (and continue to demonstrate) that in the wine sector it was possible to develop long-term entrepreneurial trajectories based on high-quality products.

At the end of the nineteenth century, the birth of cooperative wineries was the answer to excessive division of producers hindering the necessary investments to improve the productive system and the quality of the wines. Cooperatives were also used after the Second World War when the wine sector needed to overcome the damage caused by fascism (hindering the wine production in favour of table grapes) and the war (strongly reducing the surface of vines and the wine production). Cooperatives allowed (and continue to allow) small producers to have modern caves and productive system as well as an efficient retailing network. Without losing their small or medium dimension, they always remained dynamic and oriented towards an oenology more connected to the environment and traditions. They represented an alternative to the process of resource concentration made by the greatest Italian wine companies. Both the models continue to characterise the Italian wine sector in the new millennium: producers are strongly tied to their territories and, at the same time, they expanded their sales proving that the flexibility and adaptation to market trends shown by

Italian oenologists at the end of the nineteenth century are still useful today.

The fourth chapter focuses on three main aspects: (a) the changes concerning the Italian wine productive system and the related production; (b) the evolution of the presence of the Italian wine in the international markets; (c) the changes in customers' taste and consumption and the related wineries' reactions. It presents the quantitative trends of the Italian wine sector by graphs and tables: this allows to see when and how the Italian wine, particularly the bottled one, went from being a product for the domestic market to being sold in other countries. The foreign markets, in fact, played a decisive role in the modernisation of Italian wines: producers had to adapt themselves to tastes and regulations existing abroad. For example, it was particularly important to learn to sell wine in the United States, where strict hygiene laws had been promulgated for imported food stuffs since the end of the nineteenth century. The alcohol content of Italian wines was too high for the American market, so Italian producers had to create new wines. Data on export trends allow to observe the evolution of sales to US consumers. Other data on sales permit to examine the export in Germany, the top European market for Italian wines, and the current situation in the Asiatic markets. In these cases Italian wine producers' behaviours seem to copy from the past: the strategies they follow to diffuse their products in Far East foreign markets where there is no local relevant production are very similar to those used a century before in Northern Europe and South America. In this regard, the negative impact of the Covid-19 pandemic on the foreign sales of Italian wine is also considered even if data are evidently partial and moreover the emergency is not concluded.

Wine production presents evident changes over time: Italy is among the top wine producers in the world, along with France and Spain, while at the end of the nineteenth century, most of Italian wines were the result of a mixed cultivation of grapevines and other plants. The specialised viticulture started because of the fungal disease in the nineteenth century, but its diffusion became relevant in the second half of the twentieth century, in particular after the introduction of regulations concerning the protection of the denomination of origin when producers understood that the production of high-quality wines was more profitable and preferred specialised viticulture to the mixed one.

The changes concerned the consumption of Italian wine too. This last is a clear indicator of the move from a traditional, rural society to an industrialised, urban one: data show a constant decrease in the amount of wine consumed per capita. The "economic miracle" leads to dietary well-being in which the consumption of wine went from being a necessity to being a social factor and a measure of quality of life. Old inns and taverns were replaced by elegant wine bars and places to enjoy good quality wine. The old, barely drinkable "farmers' wines" became a memory of the past and new different drinks met the consumers' tastes. Sometimes, winemakers transformed their wine in a niche product with a clear identity related to the terroir: for example, in the cellars of the Alpine valleys (Valtellina) winemakers reproduced the Sfurzat brand as a "meditation wine" which "resurrected from the glorious past of mountain culture" (Locatelli and Tedeschi 2018). At the end of the twentieth century, it increased the use of the historical tradition and past as a marketing tool to sell modern wines to new consumers.

The fifth chapter exactly illustrates the construction of the image of Italian fashion also in the food and beverage sector: luxury cars, fashion, furnishing objects, and expressions of traditional Renaissance craftsmanship were transferred to Italian wines as well. These latter, as cultural products, progressively became the object of tourism trajectories towards Italy. The presence of the vineyards strongly modelled the landscape and the rural environment: the rows of vines normally got more beautiful and attractive in the countryside. The chapter verifies the actual impact of wine tourism in Italy underling that wine tourism can be established as a regional development tool: it allows the integration of the primary (agriculture), secondary (wine industry) and tertiary (tourism) sectors, highlighting the landscape attributes and showing the regional "tourism terroir" singularities. So, the tourist and wine industries are increasingly identified as natural symbiotic partners concerned with business and the sustainability of territories. This winning relationship is anchored in competitive networks, taking advantage of partnership skills and stakeholders' synergies. A complex wine tourism ecosystem is started with a deep literature review explaining that the Oenological Tourism Ecosystem includes three pillars: Wine Culture, Territory/Landscape, and Tourism. The encounter with the material and immaterial heritage is at the heart of the tourists' experience who also revaluate the culture of the Italian countryside. The findings showed a complex relational interconnectivity between stakeholders within the ecosystem, a low number of units establishing partnerships and a soft cooperation level with other players and public or private entities. So, to get business success and territorial sustainability, it is crucial that all stakeholders strengthen their partnership relations, developing strategies improving their competitive value.

The chapter also analyses the development of wine fairs from the first wine exhibitions in the nineteenth century to the main wine exposition which started during the 1960s in Verona, that is Vinitaly. The old wine exhibitions, which were exclusively reserved to producers, progressively involved consumers and new wine paths, wine cities and other big events were organised: they moved (and continue to move) thousands of tourists attracted to the "social phenomenon" of wine. Among the events organised in the oenological sector in Italy, the story of Vinitaly is briefly illustrated: it interests both producers and visitors and, moreover, represents a relevant case study. It shows the multiplicity of aspects "rotating around the current wine world", especially in Italy where wine is strictly linked to food and to global fame of Italian cuisine. As the wine panorama is another important aspect of the creation of "territorial brands" the chapter also highlights the territorial renewal policies, which are being carried out in Italy.

Furthermore the chapter briefly illustrates some notes about the economic impact of climate change on Italian viticulture. Changes in particular will concern productive yields and wine quality: this implies that prices and wine companies returns will change too. For the wine sector the most promising solutions are coming from the biodiversity and the diversification of the varieties of grapes. Climate change will permit grape growing at high altitudes in mountainous regions: what is today called "heroic viticulture" may soon be a normal and widespread winemaking practice in the valleys of the Alps and the Apennines. On the contrary, relevant problems related to the greatest climate changes will concern the wines in the Italian southern regions (in particular Sicily): the excessive alcoholic degree and the arrival of new parasites will strongly modify wine production and oblige winemakers to sell their products as blending wines or transform them in spirits and liquors.

Finally, the chapter includes a very small "addenda" including the latest wine sector trends related to the Covid-19 pandemic effects: face to the negative economic trend and the new challenges existing in the international wine markets (including the effects of the climate changes), Italian producers and public administration have to demonstrate the same flexibility and attention to innovation and quality they showed in the past (as this book illustrates).

References

Anderson, Kym. 2004. *The World's Wine Markets: Globalization at Work*. Cheltenham, UK: Edward Elgar.

Anderson, Kym, and Vicente Pinilla, eds. 2018. *Wine Globalization. A New Comparative History*. London and New York: Cambridge University Press.

Anderson, Kim, and Vicente Pinilla. 2021. Wine's Belated Globalization, 1845–2025. *Applied Economic Perspectives and Policy* 44 (2): 1–24.

Anderson, Kym, Signe Nelgen, and Vicente Pinilla, eds. 2017. *Global Wine Markets, 1860 to 2016: A Statistical Compendium*. Adelaide: University of Adelaide Press.

Biagioli, Giuliana. 2014. The Jacini Enquiry in Italy, 1877–1885. In *The Golden Age of State Enquiries. Rural Enquiries in the Nineteenth Century*, ed. Nadine Vivier, 109–128. Turnhout: Brepols.

Bodinier, Bernard, Stéphanie Lachaud, and Corinne Marache, eds. 2014. *L'univers du vin. Hommes, paysages et territoires*. Rennes: Press Universitaires de Rennes.

Cafarelli, Andrea. 2019. 'The Idea of Improvement': Theorized vs. Practised Vine Growing in Friuli. In *A History of Wine in Europe, 19th to 20th Centuries, Vol. I, Winegrowing and Regional Features*, ed. Silvia A. Conca Messina, Stephane Le Bras, Paolo Tedeschi, and Manuel Vaquero Piñeiro, 135–170. Cham: Palgrave Macmillan.

Campbell, Gwyn R., and Nathalie Guibert, eds. 2007. *Wine Society and Globalization. Multidisciplinary Perspectives on the Wine Industry*. New York: Palgrave Macmillan.

Carmona, Juan, Josep Colomé, and Juan Pan-Montojo, eds. 2001. *Viñas, bodegas y mercados. El cambio técnico en la vitivinicoltura Española (1850–1936)*. Zaragoza: Prensas Universitarias de Zaragoza.

Charters, Steve. 2006. *Wine and Society. The Social and Cultural Context of a Drink*. London: Routledge.

Cinotto, Simone. 2012. *Soft Soil, Black Grapes: The Birth of Italian Winemaking in California*. New York: New York University Press.

———. 2013. *The Italian American Table. Food, Family and Community in New York City*. Urbana: University of Illinois Press.

Colman, Tyler. 2008. *Wine Politics: How Governments, Environmentalists, Mobsters, and Critics Influence the Wines We Drink*. Berkeley: University of California Press.

Conca Messina, Silvia A. 2019. Wine Production, Markets and Institutions in Italy Between Nineteenth and Early Twentieth Centuries: A Historical Survey. In *A History of Wine in Europe, 19th to 20th Centuries, Vol. II. Markets, Trade and Regulation of Quality*, ed. Silvia A. Conca Messina, Stephane Le Bras, Paolo Tedeschi, and Manuel Vaquero Piñeiro, 177–212. Cham: Palgrave Macmillan.

Corsi, Alessandro, Eugenio Pomarici, and Roberta Sardone. 2018. Italy from 1939. Italy from 1939. In *Wine Globalization. A New Comparative History*, ed. Kym Anderson and Vicente Pinilla, 153–177. New York: Cambridge University Press.

Corsi, Alessandro, Simonetta Mazzarino, and Eugenio Pomarici. 2019. The Italian Wine Industry. In *The Palgrave Handbook of Wine Industry Economics*, ed. Adeline Alonso Ugaglia et al., 47–76. Cham: Palgrave Macmillan.

Cova, Alberto. 1988. Problemi tecnici ed economici della produzione di vino in Italia tra Otto e Novecento. In *Il vino nell'economia e nella società italiana medioevale e moderna*, 319–337. Firenze: Accademia economico-agraria dei Georgofili.

Dandolo, Francesco. 2010. *Vigneti fragili. Espansione e crisi della viticoltura nel Mezzogiorno in età liberale*. Napoli: Guida.

———. 2019. Quantity Is Not Quality: Expansion and Limits of Wine-Producing in Sicily. In *A History of Wine in Europe, 19th to 20th Centuries, Vol. II. Markets, Trade and Regulation of Quality*, ed. Silvia A. Conca Messina, Stephane Le Bras, Paolo Tedeschi, and Manuel Vaquero Piñeiro, 47–66. Cham: Palgrave Macmillan.

Dell'Osa, Dario. 2019. Viticulture and Winemaking in Abruzzo from the Unification of Italy to the Development of the Cooperation System. In *A History of Wine in Europe, 19th to 20th Centuries, Vol. I, Winegrowing and Regional Features*, ed. Silvia A. Conca Messina, Stephane Le Bras, Paolo Tedeschi, and Manuel Vaquero Piñeiro, 253–280. Cham: Palgrave Macmillan.

Federico, Giovanni, and Pablo Martinelli. 2018. Italy to 1938. In *Wine Globalization. A New Comparative History*, ed. Kym Anderson and Vicente Pinilla, 130–152. Cambridge: Cambridge University Press.

Gaeta, Davide, and Paola Corsinovi. 2014. *Economics, Governance and Politics in the Wine Market. European Union Developments*. New York: Palgrave Macmillan.

Garcie-Parpet, Marie-France. 2009. *Le marché de l'excellence: Le marché de l'excellence, les grands crus à l'épreuve de la mondialisation*. Paris: Seuil.

Hall, C. Michael, Liz Sharples, Brock Cambourne, and Niki Macionis, eds. 2004. *Wine Tourism around the World. Development, Management and Markets*. Burlington, MA: Elsevier.

Hancock, David. 2009. *Oceans of Wine. Madeira and the Emergence of American Trade and Taste*. New Haven and London: Yale University Press.

Harvey, Matt, Leanne White, and Warwick Frost, eds. 2014. *Wine and Identity. Branding, Heritage, Terroir*. London: Routledge.

Heintz, William F. 1995. *Wine Country: A History of Napa Valley: The Early Years, 1838–1920*. Santa Barbara: Capra.

Jacini, Stefano. 1884. *Relazione finale sui risultati dell'inchiesta redatta, per incarico della Giunta, dal Presidente*. Roma: Forzani e C.

Lapsley, James T. 1996. *Bottled Poetry. Napa Winemaking from Prohibition to the Modern Era*. Berkeley, Los Angeles and London: University of California Press.

Law, Marc T., and Gary D. Libecap. 2006. The Determinants of Progressive Era Reform. The Pure Food and Drugs Act of 1906. In *Corruption and Reform:*

Lessons from America's Economic History, ed. Edward L. Glaeser and Claudia Goldin, 319–342. Chicago: University of Chicago Press.

Locatelli, Andrea M., and Paolo Tedeschi. 2018. The Disappearance and Rebirth of a Wine: The Sfurzat in Valtellina. In *Des produits entre déclin et renaissance (XIVe–XXIe siècles)*, ed. Corinne Marache, Philippe Meyzie, and Maud Villeret, 71–83. Bruxelles: Peter Lang.

Loubère, Leo A. 1978. *The Red and the White. The History of Wine in France and Italy in the Nineteenth Century*. Albany: State University of New York.

———. 1990. *The Wine Revolution in France: The Twentieth Century*. Princeton, NJ: Princeton University Press.

Lukacs, P. 2012. *Inventing Wine. A New History of One of the World's Most Ancient Pleasures*. New York and London: W.W. Norton and Company.

Maffi, Luciano. 2010. *Storia di un territorio rurale. Vigne e vini nell'Oltrepò Pavese. Ambiente, società, economia*. Milan: FrancoAngeli.

———. 2012. *«Natura docens». Vignaioli e sviluppo economico dell'Oltrepò Pavese nel XIX secolo*. Milan: FrancoAngeli.

———. 2018. Il settore primario in provincia di Pavia negli anni Cinquanta. *Storia Economica* 21 (1): 157–204.

———. 2019. The Development of Wine Growing and Oenology in Southern Piedmont and Oltrepò Pavese. In *A History of Wine in Europe, 19th to 20th Centuries, Vol. I, Winegrowing and Regional Features*, ed. Silvia A. Conca Messina, Stephane Le Bras, Paolo Tedeschi, and Manuel Vaquero Piñeiro, 171–195. Cham: Palgrave Macmillan.

Maffi, Luciano, Ilaria Suffia, and Paolo Tedeschi. 2020. L'evoluzione della vitivinicoltura in Lombardia: l'Oltrepò Pavese, la Valtellina e la Franciacorta nella seconda metà del Novecento. In *Il settore agro-alimentare nella storia dell'economia europea*, ed. Giovanni Gregorini and Ezio Ritrovato, 213–234. Milan: FrancoAngeli.

Maffi, Luciano, Paolo Tedeschi, and Manuel Vaquero Piñeiro. 2021a. A New Wine for the International Market: Italian Public Institutions' Initiatives Supporting the Oenological Sector (1870–1910). In *Wine, Networks and Scales. Intermediation in the Production, Distribution and Consumption of Wine*, ed. Stéphanie Lachaud-Martin, Corinne Marache, Julie McIntyre, and Mikaël Pierre, 115–130. Bruxelles: Peter Lang.

———. 2021b. Winemakers and Wine Industry Developments in Italy (1884–1914). *Journal of Wine Research* 32 (4): 229–246.

Mocarelli, Luca, and Manuel Vaquero Piñeiro. 2019. Viniculture in the Italy of the Mezzadria (Tuscany, Umbria and Marche). In *A History of Wine in Europe, 19th to 20th Centuries, Vol. I, Winegrowing and Regional Features*, ed. Silvia A. Conca Messina, Stephane Le Bras, Paolo Tedeschi, and Manuel Vaquero Piñeiro, 227–252. Cham: Palgrave Macmillan.

Oddy, Derek J., and Alain Drouard, eds. 2016. *The Food Industries of Europe in the Nineteenth and Twentieth Centuries.* London and New York: Routledge.

Pan-Montojo, Juan. 1994. *La bodega del mundo. La vid y el vino en España (1800–1936).* Madrid: Alianza.

Paul, Harry W. 1996. *Science, Vine and Wine in Modern France.* Cambridge: CUP.

Pedrocco, Giorgio. 1994. Un caso e un modello: viticoltura ed industria enologica. In *Studi sull'agricoltura italiana: società rurale e modernizzazione*, ed. Pier Paolo D'Attorre and Alberto De Bernardi, 315–342. Milano: Feltrinelli.

———. 2000. Viticoltura ed enologia in Italia nel XIX secolo. In *La vite e il vino: storia e diritto (secoli XI–XIX)*, ed. Mario Da Passano et al., 613–627. Roma: Carocci.

Rapetti, Vittorio. 1984. *Uomini, collina e vigneto in Piemonte, da metà Ottocento agli anni Trenta.* Alessandria: Edizioni dell'Orso.

Ritrovato, Ezio. 2010. Sviluppo dei mercati vinicoli dopo il 1860. In *Storia regionale della vite e del vino. Le Puglie*, ed. Antonio Calò and Liana Bertoldi Lenoci, 531–550. Martina Franca: Edizioni Pugliesi.

———. 2019. The Wines of Apulia: The Creation of a Regional Brand. In *A History of Wine in Europe, 19th to 20th Centuries, Vol. II. Markets, Trade and Regulation of Quality*, ed. Silvia A. Conca Messina, Stephane Le Bras, Paolo Tedeschi, and Manuel Vaquero Piñeiro, 117–136. Cham: Palgrave Macmillan.

Simpson, James. 2011. *Creating Wine. The Emergence of a World Industry, 1840–1914.* Princeton and Oxford: Princeton University Press.

Stanziani, Alessandro, ed. 2004. *La qualité des produits en France, XVIIIe–XXe siècle.* Paris: Belin.

Tedeschi, Paolo. 2008. Marché foncier et systèmes de production agricoles dans l'Italie du nord au XIXe siècle: le cas de la Lombardie orientale. *European Review of History / Revue Européenne d'Histoire*, no. 5: 459–477.

———. 2017. Note di eno-storia economica: viticoltura e produzione di vino nei terroirs italiani (secoli XVIII–XX). *Proposte e ricerche*, no. 79: 101–111.

———. 2019. The Improvement of the Production and Quality: The Case of Wine Production in the Eastern Lombardy During the Nineteenth and Twentieth Centuries (Provinces of Bergamo and Brescia). In *A History of Wine in Europe, 19th to 20th Centuries, Vol. I, Winegrowing and Regional Features*, ed. Silvia A. Conca Messina, Stephane Le Bras, Paolo Tedeschi, and Manuel Vaquero Piñeiro, 197–226. Cham: Palgrave Macmillan.

Teughels, Nelleke, and Peter Scholliers, eds. 2017. *A Taste of Progress: Food at International and World Exhibitions in the Nineteenth and Twentieth Centuries.* London and New York: Routledge.

Unwin, Tim. 1996. *Wine and the Vine: An Historical Geography of Viticulture and the Wine Trade.* London and New York: Routledge.

Vaquero Piñeiro, Manuel. 2012. *Storia regionale della vite e del vino in Italia. Umbria*. Perugia: Volumnia.

Zaninelli, Sergio. 1977. Un tema di storia dell'agricoltura italiana fra Otto e Novecento: la diffusione della fillossera ed il rinnovamento della viticoltura. In *Fatti e idee di storia economica nei secoli XII–XX*, ed. Carlo M. Cipolla et al., 861–878. Bologna: Il Mulino.

CHAPTER 2

Institutional Innovation and Public Policies

The evolution of public policies in favour of the development of the agricultural sector represents a classic research topic, on which a wide historical literature is available (Tracy 1989). Main theses establish an evolutionary path running from the second half of the nineteenth century to the late twentieth century and beyond: public intervention in the wine sector was relevant and helped the modernisation of the winegrowing and winemaking, a concept that since the nineteenth century became almost synonymous with industrialisation (Moser and Varley 2013; Vivier 2008). The development of Italian viticulture provides a clear example of the interaction between the State and agricultural modernisation: the public intervention was in fact necessary to initiate a decisive industrialisation of the Italian primary sector.

Viticulture was present throughout the peninsula and covered about 17 per cent of the available agricultural land. From the 1870s onwards, the State began to take vigorous action to improve Italian winegrowing. While in France winegrowers' associations became a force of political pressure to counter the depressive effects of falling prices (Planas 2016), in Italy the State pushed for a reforming action: it improved the school system, encouraged cooperatives, spread technology, internationalised the market, and promulgated laws against fraud and manipulation of drinks. After the First World War, Fascism played an important role in legislation and in the debate on typical wines: however, it did not help the development of the oenological sector. Finally, the entry into force of European

M. Vaquero Piñeiro et al., *A History of Italian Wine*,
https://doi.org/10.1007/978-3-031-06097-7_2

Economic Community regulations enlarged the public intervention concerning Italian viticulture and made a relevant important phase of transformation with the law on designations of origin: the relationship between winegrowing and the landscape increased, and a new wine culture was created.

2.1 Education and Professional Training: Schools of Oenology

The European countryside in the nineteenth century was characterised by an intense educational offensive (Van Molle 2005; Depecker and Joly 2015). The European landscape was rich in initiatives and figures who carried out energetic action to improve the educational condition of the peasant masses. From various sides, creating a unique patrimony of public and private ideas, the promoters of the renewal of the educational bases of continental agriculture dedicated themselves to consolidating progress through science and technology. Overall, a supranational context emerged that was favourable to the intervention of scientists and technicians as the bearers of modernity, generating a mechanism for the dissemination of knowledge passing from the top to the lowest levels of rural society. The consequence was the formalisation of an educational chain going from universities and specialised research centres to the lower levels, thanks to the integration of practical schools, elementary schools, courses, and clubs (Rösener 1994; Locatelli and Tedeschi 2015, 2018). In a scenario dominated by deep-rooted private interests and consolidated social differences, the State called upon during the nineteenth century to stimulate an economic modernisation which, for many members of the élites, was not to undermine the foundations of the social order (Prosperi 2019).

From the mid-nineteenth century onwards, there was an intense dialectic between landowners who did not accept changes and state forces committed to giving agriculture a more dynamic structure. For European states, stimulating agrarian progress was a matter of national interest (Moser and Varley 2013): European countries took the lead in institutionalising and disseminating new thinking to be transferred to agricultural production. Agrarian issues became part of the national-political debate, and late nineteenth-century agriculture became increasingly complex and rich in professional paths. Teachers, dealers in machinery and chemical fertilisers, agronomists, farm managers, and ministerial officials made agriculture a suitable space for social and economic growth (Betri 2002). The land continued to confer honour and prestige on people: at the same time, the industrial revolution and the expansion of international trade gave a space to sectors of the rural population seeking visibility and their own sphere of affirmation (Chambers and Mingay 1969). While higher education in all countries remained very elitist (Malatesta 2006), the state was committed to the diversification of education through the creation of study centres for technical and applied disciplines. While in England formal agricultural education advanced slowly (Brassley 2000, 2005; Beardmore 2015, 2019; Bowen 2010; Webster 2007; Hurren 2016), on the European continent, state initiatives in the field of practical education proliferated from the early nineteenth century, intended to change the patterns that rigidly conditioned the countryside (Institut International d'Agriculture 1936; Segers and Hermans 2009; Borras 2009; Jas 1997; Cartaña i Pinén 2000; Pan-Montojo 1995; Gabriel (de) Fernández 1963; Cornu and Mayaud 2007; Denis 2013; Harms et al. 2013; Forney 2011; Harwood 2005).

Giving relevance to the relationship between schooling, social change, scientific progress, and the modernisation of agriculture (Philippe and Kingsland 2015), this chapter analyses the special schools established in Italy in the late nineteenth century with the aim of training technical and managerial personnel in viticulture and oenology. Italian State in fact created a geographically wide public educational offer allowing to achieve a strong modernisation of the wine sector.

The analysis is divided into two parts. The first concerns the period from the end of the Napoleonic age to the 1870s: private initiatives in the field of agricultural education began in Italy and the most famous case was represented by the Meleto school founded in Tuscany by Marquis Cosimo Ridolfi. It constituted a first attempt to introduce new subjects and new teaching methods in Italy, while maintaining the rules regarding production and social relations in the countryside. Meleto's experience, much praised, was an isolated example because the élites (that is the Saxon genitive) thought that it was enough to learn by example and practice, leaving aside any attempt at technical and scientific renewal: besides, they considered not useful to spend money to enlarge the know-how of peasants having no possibilities to own land where apply their knowledge. The second part of the analysis presents the following evolution after the 1870s when the State created a public teaching system specifically designed to train professionals for agriculture in general and viticulture in particular. So, the Italian case constitutes an example of the reforming action of the State through new education and modernisation processes in the countryside before the First World War as it also happened in France (Weber 1976; Phillips 2016).

Financed by public administration, the special schools of viticulture and oenology were organised since 1877, along the lines of similar institutions in France and Prussia (Pazzagli 2005). They were intended to encourage the emergence in Italy of a modern workforce of winery managers, technicians, and modern smallholders. Italian State pushed the creation of formal training courses that guaranteed access to professions linked to the administration of farms. Students from the special schools of oenology and viticulture after a regulated training course sanctioned by the granting of a diploma contributed decisively to the professionalisation of Italian agriculture (Perkin 1989, 1996). They formed a specialised, diversified, and qualified workforce willing to claim status, an income, and the conviction that their services were essential for the Italian society's progress.

When the Kingdom of Italy was born in 1861, agronomic knowledge was the fruit of a plurality of public and private institutions (municipalities, religious bodies, academies, philanthropic circles) (Stringher 1911; Giglioli 1903; Zaninelli 1990; Vaquero Piñeiro 2015; Govoni 2011; Failla and Fumi 2006; Bidolli 2001). The initiatives were the result of a spirit that could be traced back to the Enlightenment of the eighteenth century (Jones 2016): in many circumstances, the result was a combination of notions concerning botany, veterinary science, geology, and other related subjects (Ambrosoli 1997). Despite the dominance of a rather uneven agronomic landscape (Pazzagli 2008), in the first half of the nineteenth century an agronomic network was built and overcame the barriers imposed by political boundaries (Locatelli and Tedeschi 2015).

As an expression of aristocratic reformism, the foundation in 1834 of the school in Meleto (Tuscany) by the Marquis Cosimo Ridolfi was an initiative of considerable importance (Pazzagli 2004). Marquis Ridolfi was determined to put into practice on his farms what he had learnt during his travels around Europe: his school initially stimulated interest because of the pedagogical orientation adopted (Mirri 2004). In a few years, Meleto Castle became a model for the large landowners who were favourable to the introduction of innovations in agriculture as long as they did not favour changes in the existing production structures and relations. Landowners appreciating Ridolfi's work: it was necessary to carry out a technical modernisation, avoiding the danger of creating the conditions for the formation of a social climate leading to the spread of changes considered too radical. This consolidated one of the main characteristics of Italian agriculture during the nineteenth century: the prevalence of a purely technocratic approach to the modernisation of the rural sector.

Meleto's experience was short-lived because in 1842 it ceased to function despite having encouraged the formation of similar educational institutions in other Italian towns (Jesi, Modena, Pesaro, Alessandria). This initiative in the field of training qualified personnel to be entrusted with the tasks of farm management played a crucial role: it in fact allowed to build a favourable climate for the technology transfer in agriculture (Mazzotti 2017; Petrusewicz 1991).

In the same years, Baron Bettino Ricasoli also began his research to obtain the "perfect Italian wine" (Ciuffoletti 2009). Guided by science and the renewal of agriculture through technical experimentation (Dewey 2013), he transformed his Brolio castle into a sort of oenological laboratory where Chianti became the "true type of red wine for meals, as [...] today's consumer tastes and the demands of the wine trade demand" (Ciuffoletti 2009). Ricasoli's achievements reinforced the idea that Italian winemakers needed to learn how to "govern wines in the Tuscan way": even if few winemakers had the right skills to achieve the quality levels of Brolio's Chianti, there was a widespread opinion that it was possible to obtain accredited table wines, putting an end to the custom of selling off poor-quality wines. Chianti became synonymous with "Tuscan-style red wine" regardless of its place of origin and characteristics. This also explains why during the twentieth century the official delimitation of the Chianti area, existing since the eighteenth century, provoked harsh disputes and controversies (Meloni and Swinnen 2018).

After the birth of the Kingdom of Italy, there were no significant changes concerning school and education in the countryside. For many great landowners, the state should not intervene in agriculture: this last was in fact conceived as the logical result of the free initiative of owners reaching an agreement with the sharecroppers or peasants for the exploitation of the property land. Any initiative promoted by the government was negatively judged because it could jeopardise the relationships regulating the agrarian contracts. In the absence of real liberal proposals, the first attempts to introduce a system of training in the field of agricultural knowledge in Italy ended up in an absolute mess (Ivone 1982).

Face to landowners who were against state intervention, there was a growing pressure from the landowners who wanted to modernise the sector, including the creation of a school system capable of meeting the challenges imposed by the capitalist transformation of the world agriculture (Cecil 1979). What was at stake was Italy's position in the international division of labour: élites had to choose either among the countries producing raw materials and essential goods or among the industrial powers. The problems imposed by economic development fuelled the national political debate on the need to have a school system geared towards the technical training of personnel employed in agriculture. Agronomic knowledge could not remain in the philanthropic élites' hands (Vivier 2008, 2009).

When the landowner lived in the city and visited his property a few times a year, the management of farms and wineries was the task of the "fattore" or director, a skilled worker who stood between the owners and the sharecroppers and peasants. He learnt through experience and direct practice: he also had little schooling and was the result of an agricultural system based on the full control of peasant labour. However, at the end of the nineteenth century, he began to prove unsuitable because it became a priority to have knowledge of mechanics, chemistry, hydraulics and other technical-scientific subjects to be used in the management of modern farms and cellars, moreover when vineyards were being struck by phylloxera or fungal diseases.

Between 1870 and 1872, the high schools of Portici (Naples) and Milan came into operation: they were in addition to the old one existing in Pisa. However, the main novelty came in 1879 when the Italian parliament, overcoming resistances concerning the establishment of the "public general education", approved the law imposing the creation of practical and special agricultural schools (Cheneaux-Berthelo 2020). A double order of agricultural schools was created: the general theoretical-practical teaching of agriculture was entrusted to the practical agricultural schools while the teaching concerning some particular agricultural sectors was the task of specific special schools (viticulture and oenology; fruit growing; zootechnics and dairy farming).

Studying the modernisation of the wine-growing sector means identifying the channels through which innovations were disseminated: the special oenology schools founded by the Italian State played a fundamental role. The founding of a school system for training agricultural technicians initially led to criticism and heated debate regarding the special schools for viticulture and oenology. Among those who were opposed to broadening the scope of agricultural knowledge, it was argued that peasant students were ill-suited to scientific instruction. They thought that educational institutions aimed at training specialists in the field of viticulture and oenology would not be successful, because the owners not only had little agronomic knowledge, but also paid little attention to academic advice.

They considered that agricultural practice was learned by doing, by experience and by following the advice of older people. According to the critics, before creating highly selective educational courses, such as those on the grape and wine cycle, it was necessary to raise the general cultural level through lectures, useful reading and easy courses. Only at a later stage, and once a certain basic culture had been acquired, wine schools could become a useful and effective tool. These opinions were contrasted by the most enterprising sectors of Italian wine-growing: they wanted to operate in the competitive international context, and they were in favour of new schools for winemakers. They thought that only by following the dictates of science and technical knowledge it was possible to learn how to combat the diseases that affected plants, how to use the new tools for winemaking and generally improve the quality of wines.

The creation of a specific school for wine education fuelled strong tensions. For many landowners, the traditional relations of power and production were threatened by a schooling centred on entrepreneurs, technicians, and wage earners and all oriented in a capitalist way to the market and profit (Bidolli and Soldani 2001). The model followed by the public administration was represented by France and Prussia, where the foundation of specialised institutions allowed the respective national oenologies to gain important market shares: at the end of the nineteenth century, only wines from Moselle, Rhineland, Burgundy, and Champagne were in fact on the menus of the world's best restaurants. Italian wines were absent: only Chianti and Asti spumante were known abroad.

The context started to change from the end of the 1870s as a result of the "agricultural crisis" and the changes that were taking place in world trade. While the reduction in cereal prices increased the purchasing power of nominal wages, it also reduced the labour supply and provoked for a large section of rural society a relevant deterioration in living conditions in the Italian countryside: this led to strikes. Cereal production was reduced in favour of vines, olives, or livestock farming requiring a higher level of workers' professional skills. The new agrarian school system created new jobs for graduated students as well as opportunities for socio-economic growth. New special schools of oenology had the aim to prepare directors of wineries and experimental cellars as well as technical and specialised personnel for the public administration and associations against the phylloxera. The Italian oenology schools became the model for non-European countries such as Argentina, which were engaged in the construction of national wine sectors. State created schools in Conegliano, Alba,

Avellino, Catania, and Cagliari. In addition, the Royal Experimental Oenological Station was created in Alba and had a research function. In 1923, another experimental station in Conegliano became operative. In the context of growing attention to the technical and scientific aspects of wine-growing, a decisive turning point was represented by the diffusion of phylloxera and fungal diseases that required a rapid change in winegrowers' mentality, abandoning traditional methods and practices to make wine production a truly industrial activity.

The educational orientation of the special schools of viticulture and oenology was adapted to the characteristics of each region. The training course was divided into two levels of theoretical-practical teaching related to the cultivation of the vine and the production of wine in order to train people suitable for teaching the science, managing companies and oenological societies and the practical exercise of vineyard activities and the preparation and conservation of wines. At the end of the first training cycle, the title of head winemaker or site manager was obtained, while the second training period ended with the graduation of "wine technician".

The public school system favoured the formation of a rural society with greater professional diversification. Between landowners and sharecroppers and peasants, there were new workers seizing the opportunities offered by the increasing modernisation of the agriculture. Students attending university courses were a minority. The higher number of students in practical and special schools put in evidence that the technical renewal linked to agriculture was essentially at the intermediate levels of the school-education system. The main objective of the new agricultural education was to create a solid base of technicians and administrators who contributed to the modernisation of Italian farms and wineries.

The results of the massive deployment of public means and resources to improve the level of knowledge in agriculture can be seen by observing the professional activities carried out by the young people who graduated from the Royal Oenology Schools. According to the data available for the Conegliano school, by 1884 there were 52 students who had completed their training: most of them (31) found management and responsibility jobs in some wineries or oenological establishments; another group took up teaching abroad (e.g. in Buenos Aires), while others were employed in family companies or specialised in the international wine trade. Special mention for the oenologists who, after leaving the oenology schools, moved on to prestigious positions in the public administration, such as the direction of the oenotechnical stations founded by the Italian government in the main European and American towns (London, New York, Buenos Aires, Munich). In the following years, during the period 1885–1893, there were 309 students who graduated after three years of study: 73 (24 per cent) found a job as vine-dressers or specialised workers; 87 (28 per cent) returned home to devote themselves to the management of the family land properties, putting into practice what they had learnt during their school years; 59 (20 per cent) continued their studies, especially in higher education courses in oenology; 31 (10 per cent) began a self-employed profession connected with agriculture, and 59 (20 per cent) appear in the statistics without a specific occupation (MAIC 1894).

At the end of the nineteenth century, numerous socio-professional figures in fact tended to escape official statistics: this also explains why 20 per cent of graduated did not have a job. In the censuses, many professions were not easy to define, and there was no detailed information on studies carried out or on the possession of a school graduation. Even if available data were very scarce, the presence in the censuses after 1900 of the category of wine technicians working in winemaking plants confirmed the strengthening of technical figures in agriculture. They represented a minority group, but the data showed that new education system in agriculture had positive results. The path to employment for students who graduated from the oenology schools was neither easy nor immediate. They in fact had to contend with the resistance of the old directors of the wine cellars, but they also had overcome common practices for the cellars: for example, as it happened in Sicily, these latter were directly governed by owners thinking that the oenology schools had to limit themselves to training good workers (vine-dressers), renouncing the objective of training real directors of wineries (MAIC 1894).

Despite resistance, the modernisation in Italian agriculture consolidated the presence of qualified workers who were able to foster oenological progresses (Augello and Guidi 2012): this was helped by the new law of 1900 approving the staffing of wine technicians both at home and abroad, as well as the directors and assistants of the experimental cellars (MAIC 1900).

Concerning the historiographical debate on the role played by the state in changing the economic paradigm through investment in education and training, empirical evidence shows that at the end of the nineteenth century government action was essential (Luque Ballesteros 2006). In Italy new public educational offer in the agricultural sector gave the possibility to introduce numerous elements of novelty and change. Thanks to the formation of an extensive network of schools and institutions distributed from Northern to Southern Italy, the rural population had the opportunity to improve their quality of life. Without reaching the level of university education, the practical and special agricultural schools gave many young people, coming from farming classes, the chance to obtain a professional specialisation: so, some rigid patterns and hierarchies began to fall away.

The general framework, despite the weight of the school system, oscillated between the consolidated strength of tradition and the need to incorporate the knowledge of young technical graduates who began to push for having their title recognised as a priority qualification for occupying management positions (Cazzola 1996). The dispute between those who had learned from their elders only and the new technicians trained in the new schools, having much more theory than practice, characterised the start of the real modernisation of Italian agriculture and obviously the Italian wine sector. While the old farm managers had consolidated experience, the young people proved to be superior when it came to making plans, measurements, designing hydraulic works, fighting the diseases that affected the plants, and much more. In this way, during the conflicts agitating the Italian countryside at the beginning of the twentieth century (Caron and Chauvaud 2005), the rural modernisation based on the recourse to the technology became the main instrument to develop the quality and quantity of harvests (including the grapes ones). Through the action of the royal practical and special agronomic schools, public administration was not interested in offering peasants the possibility of specialised studies, the aim was the formation in the country of a body of technicians bound to the practical reality of agriculture. A technocratic vision, tending to be conservative, destined to furtherly consolidate after the 1920s with the arrival of fascism (Fernández-Prieto et al. 2014) (Image 2.1).

At the beginning of the twentieth century, the main change was represented by the arrival on the labour market of graduates in practical and special schools (Vaquero Piñeiro 2011, 2016). Young people who had acquired a technical education and training in agriculture started to put what they had learnt into practice and gained recognition in terms of career progression and income. This generated bitter conflict among technicians in agriculture. Some graduates went into technical professions related to agriculture such as surveying, machine building, and measuring.

Image 2.1 *Remuage* of bottles of sparkling wine carried out by students of agricultural schools (1930s). (Source: Archivio Storico Indire, Fondo Fotografico)

Others found employment in public administration. Others found employment in the central and provincial public administration as professors, managers of land reclamation and agrarian consortia, wine technicians in embassies, and so on. The largest group worked as directors in agricultural companies and factories. A clear distinction was finally created between the young graduates who had come out of school after years of study and the old factors who had not studied but retained a strong power.

2.2 Technological Modernisation: Between Public Intervention and Private Initiative

Technology has played a relevant role in the development of the Italian wine sector. The scientific literature confirms the complementarity between new technology and economic development (Kookson 2018; Macedo and

Valentines Álvarez 2016; Mokyr 1990). Many studies related the role played by technological progress in global economic leadership. In the second half of the nineteenth century, the mechanistic-technical modernisation resulting from scientific discoveries and industrial progress involved all areas of production, including agriculture (Moe 2007; Timmons 2005; Onger 2010). The study of the leverage of technology in the primary sector pushed investigations in the direction of products such as cereals or seeds, while others, such as oenology, had less historiographic success even if they had, in some regions, the same economic and social relevance. Besides, the changes concerning viticulture and winemaking during the nineteenth century represented very interesting examples of the effects provoked by the spread of an environment conducive to the standardisation of production processes using new technologies (Simpson 2011).

Among the external stimuli that brought about strong changes in the wine sector, it is important to underline the spread of destructive plagues. The cryptogamic diseases of the vine (downy mildew, powdery mildew) and the phylloxera, which struck most European countries almost at the same time (Paul 1996), obliged the wine sector to undertake a difficult and challenging cultural modernisation (Ainsworth 1981; Egerton 2012; Hernández Adell and Pujol-Andreu 2016). It was necessary to definitively abandon old and outdated beliefs and this created a distinction between small-scale farmers, who continued to produce for their own consumption or the local market using traditional methods, and new wineries investing their capitals (collected among their members, sometimes gathered in cooperative societies) for the improvement of the winegrowing and winemaking as well as for the modernisation of storage facilities (Unwin 1991).

The diffusion of the agronomic knowledge and new chemical products helped wine sector to strongly reduce the effects of plant diseases: however, the recourse to the sulphurisation and Bordeaux mixture increased the production costs and progressively reduced the surfaces for viticulture. It was in fact profitable to maintain vines only where the yields in grapes and the quality of the wine allowed to compensate for the new costs. This aspect became more relevant at the end of the nineteenth century: to eliminate the destruction provoked by the phylloxera, it was necessary to graft all autoctones wines with American ones and the related high cost obliged the intervention of the public administration to support viticulturists. All this obviously implied a change of the way of understanding and practising wine-growing as well as a renewal of working tools to improve the quality (and the price) of new wines (Grübler 2003).

The old storage cellars were transformed into industrial winemaking plants equipped with new technology and designed according to precise engineering criteria: this evidently increased the weight of technological input in the Italian wine sector. At the end of the nineteenth century, if compared to France and the new main wine-growing countries entering the world wine market (as United States, Argentina, Australia, South Africa), Italian wine sector had a low inclination towards renewal: however, it opened up to the mechanisation of production processes (Kookson 2018; Timmons 2005; Henzell 2007) and it benefited from the enlargement of world wine trades (Aparicio et al. 2009) where it was also dragged by the great results of other Italian agri-food products in the new global markets (Nützenadel 2008; Stanziani 2005; Pilcher 2006). The technological modernisation of Italian oenology was slow due to the existence of a relevant domestic demand that guaranteed the sale of large quantities of poor-quality wine. So, most of small producers were not interested in technological progress: someone also argued incorrect use of new chemical products. New technology was instead one of the tools allowing the best Italian winemakers to make a qualitative leap in their products and progressively became famous and appreciated in the international markets.

In Italian wine sector manual operations continued to predominate and caused a lack of uniformity concerning wine realised in different seasons: this did not depend on different climate conditions of the year, but also by the existing limits in winemaking and in the old methods concerning the conservation of the wort and wine. The change began when the pages of specialised newspapers and manuals started to write about the need to obtain "constant" and "typical" wines: this was the starting point for a standardisation of production processes, an essential condition for the international marketing of bottled quality wines (Stanziani 2004). In best wineries, the modernisation involved the entire wine production chain, from the cultivation and care of the plants to the last step of marketing, passing through vinification and storage in the new cellars. Thanks to the progressive mechanisation of production processes Italy was able to make, in the early twentieth century, an important breakthrough on international markets, replacing exports of cheap blending wines, used to "cut" foreign wines which needed to increase their alcohol content, with more refined wines which were included on restaurant menus in the world's major cities (Conca Messina et al. 2019).

The technological transformation of the Italian wine sector was reinforced by the intervention of the State, which drew up a strategy for the types of wine to be targeted. This became more evident in 1885, when a competition was launched to reward the production of table wine. Favouring table wine, to the exclusion of the much easier to obtain wine for blending, was a precise choice of agrarian policy, because this type of wine required oenological skills and suitable equipment (Federico and Martinelli 2018; Vaquero Piñeiro and Maffi 2018). Professor Michele Carlucci's final report revealed a situation made up of shadows and lights: he first highlighted the importance of the wine sector for the Italian economy, as evidenced by the growth of both wine production between 1870 and 1883 and the related positive trend of exports (MAIC 1884). However, he illustrated that the increase was largely made up of "raw wines", that is wines known as "for blending or cutting" (Carlucci 1890). This indicated that in Italy the production of table wines, although growing, remained in the minority of wineries.

In order to verify the recommendations of the oenological books, the public members of the jury evaluating wines for the 1885 award emphasised that the latest equipment had to appear adequate for modern winemaking, as it allowed a fast and constant production chain. They wished that the recourse to technological innovations became normal for best wineries whose aim was to produce and trade in table wines and sparkling wines. Their report also indicated that there existed huge investments by wine entrepreneurs in technological updating, a clear sign of the rise of Italian wine capitalism.

Screw presses were among the equipment registered in almost all wineries. Their use became more widespread in the second half of the nineteenth century in cellar practices allowing a new relevant progress in the Italian winemaking. The process of filtering wine was an established practice and each winery had suitable equipment according to the most widespread commercial offer (Dutch filters, Carpené filters, Rouhette). Some winemakers' entrepreneurial spirit was evidently underlined by the presence of machinery imported from abroad specifically for the production of wines requiring advanced processes: among them emerged the "English steam-powered machine for the production of sparkling wines". So, in best Italian wineries it was possible to observe the presence of corking machines (including a Thiabaud), cork markers, capsuling machines, bottle washers, and oenological pumps.

Numerous instruments were used for chemical and physical analysis and helped the technological modernisation of Italian winemaking plants: Malligand oenometers, ebulliometers, Houdart oenobarometers, acidometers, glucometers, and alcoholometers. Tools and machinery were essential for the correct handling of the raw material, but members of the ministerial commission also gave more attention towards controlling the quality of the wine vessels used. Alongside oak vats and barrels, other novelties were associated with the presence in the wineries of large concrete tanks internally lined with glass tiles. Italian wine cellars visited by the jury were in the vanguard: large masonry cisterns with a capacity of 200–300 hectolitres allowed large quantities of wine to be stored in optimum conditions while taking up little space.

The competition of 1885 showed that one of the means adopted by the Italian State to encourage the mechanical and technical modernisation of the wine sector was the granting of premiums. Another public strategy was the creation of government warehouses of machines that farmers and owners could borrow when they were necessary. The depots were set up at the royal practical agricultural schools, which allowed winemakers the opportunity to get in touch with the most advanced processes in the grape handling sector (Bidolli and Soldani 2001).

The technological progress made by the Italian wine sector was also evident in the advertising pages of newspapers and specialised trade guides (Circolo enofilo italiano 1892). The volumes, accompanied by advertisements for companies operating in the wine sector, well represent the effects of the increase in demand and supply of oenological equipment. This vitality also explained the presence of national equipment agencies from France, Germany and England: the company L. Rinaldi & C. of Rome had the exclusive representative in Italy of the Clayton & Shuttleworth company of Lincoln (England); the company C. Malenchini & C. of Florence represented for central Italy the English company Ruston & C. (from Lincoln too). Messages also concerned factories directly active abroad, such as Giulio Roller and Carl Bartelt, both in Frankfurt am Main (Germany). In towns such as Bologna, Milan, Turin, and Florence, new industries specialising in the manufacture of winemaking machines were set up: they used foreign patents that had been appropriately modified. One sector experiencing a real revolution was the construction of tank wagons and containers specifically for transporting wine by train. One of the most famous companies was Garolla & Limena of Padua: at the end of the nineteenth century became a renowned mechanical factory specialising in the production of crushers, stemmers, and centrifuges. The trading house of Ottavio Ottavi was relevant too: in addition to selling tools and chemical fertilisers, Ottavi founded in 1875 the *Giornale vinicolo italiano* (literally "Italian wine journal") and wrote numerous volumes on the rational practice of viticulture. In particular, his monograph ***Enologia teorica-pratica*** ("Theoretical-Practical Oenology") had 11 editions and became the reference text on Italian oenology in the transition from the nineteenth century to the twentieth.

Californian wineries became the object of visits and study in order to learn about organisational systems and also technical solutions suitable for saving time and rationalising manual work, especially after the arrival of steam engines (Vaquero Piñeiro and Maffi 2018). For Italian viticulture, news arriving from the United States about the construction of large plants equipped with modern technology represented a relevant point of comparison. Press articles and official reports highlighted the progress made by wineries in California. It was clear that this was a wine industry that could become a threat to the countries of Mediterranean Europe (Alston et al. 2018). In particular, the architectural solutions used in the construction of the cellars were admired and imitated. So, following the most innovative trends, a new type of oenological establishments designed by engineers and conceived with the extensive use of machines were built at the end of the nineteenth century.

With the architectural transformation of the buildings, there was a growing awareness of the need for oenological machines that were more and more suitable for the production of quality wine. This further transformation process was based on two symbolic machines: the presses and the crushers. The best wine, of course, was obtained from the pressing of grapes and the subsequent fermentation of the must. However, many wineries which wanted to increase their production without thinking about the quality, made other operations: the pressing of the marc, with or without the addition of water, in fact granted a low-quality secondary product.

Most of winemakers' production methods were old: the pressing continued to be carried out in the open air with fermentation in vats or tubs left uncovered, after which the must was placed in storage barrels, usually in damp, mould-filled cellars. Under these conditions, the wine did not last more than a year. Besides, after the first pressure, other pressing operations began, in order to obtain another wine. Hot water was also poured over the already pressed marc: this last, combined with the remaining liquid, formed a third wine, which was used by the daily labourers and for the family's common use (Maffi 2012).

The first real mechanical crusher was created by Dr Ignazio Lomeni and presented in Milan in 1823 (Ottavi and Strucchi 1908; Marescalchi and Dalmasso 1979). Two years later, a Lomeni's pamphlet was published on a "machine for crushing grapes", which was awarded a silver medal by the Imperial Regno Istituto di Scienze, Lettere ed Arti of Milan (Lomeni 1825). In 1837, at the Milan Exhibition, the carpenter and mechanic Giuseppe Torti from Cologne (Eastern Lombardy) was awarded a silver medal for "a mechanism for crushing grapes and separating the berries from the stalks". The machine was granted an exclusive five-year privilege starting on 25 March 1837. The following year, Giandomenico Silva (also coming from Eastern Lombardy) presented a grape crusher at the extraordinary exhibition in Milan, as evidence of the Lombard agronomists' inventiveness (Onger 2010).

This oenological practice favoured the quality of wines: few years later a further improvement came with the grape crushers which, by favouring the immediate separation of the stalks, allowed less extraction of unpleasant tannic substances. As described by Felice Garelli, the "mechanical mashers or crushers" were essentially made up of two cylinders with slats that could be moved apart, covered with a mesh of galvanised iron wires, placed under a hopper, and operated mechanically by a crank. The experimentation of machines for extracting must using centrifugal force could improve the quality of wines, but no authors expressed a judgement on the actual effectiveness of that system (Garelli 1873; Ottavi and Strucchi 1908; Guffanti 1877).

In the second half of the nineteenth century, crushing-grinding machines were progressively diffused in all large processing cellars: many models were proposed to winemakers by the wine industry. Those considered avant-garde carried out the stripping of the grapes and the pressing at the same time, making the two operations simultaneous and saving time and labour. One of the best models was made by Bruggemann in Germany: it consisted of a rather narrow circular drum arranged vertically, with a horizontal axis with six blades rotating in the centre of the drum. The lower part of the drum's circular walls was made up of a grid, which provided a passageway for the grapes and must. It was quick and economical to operate. Its cost was estimated 15 Italian liras per 100 quintals of grapes and its price was 280 Italian liras.

One of the most powerful pieces of equipment built to serve large-scale industry was the centrifugal crusher-grinder manufactured by Garolla in Limena. It was equipped with a hollow cylinder formed by an iron frame, covered with tinned metal cloth. Concentric to this first cylindrical sieve was another one which was made of smaller mesh. The price varied from

200 to 300 Italian liras for those that could be moved by hand and it arrived up to 2000 Italian liras for those powered by steam (Ravizza 1900; Ottavi 1898, 1900).

Concerning foreign machinery which was distributed on the Italian market, the main presses were made by V. Coq of Aix-en-Provence, E. Françon of Marseilles, and Simon et fils of Cherbourg. English technology was also used, as demonstrated by the presence of Bale & Edwards in Milan and Naples: this company was active from 1876 to 1902 and allowed Italian viticulturists to know the most advanced technologies of the English wine industry (Bigatti 2000). In communicating the characteristics of its machines to gain the trust of Italian winemakers, it wrote that the presses, apart from their speed, were reliable for any inexperienced person. The engineer Alfred Ormond Edwards was appointed vice-consul of the United Kingdom and networked with the most important Italian entrepreneurs, including Giovanni Battista Pirelli.

In 1903, Professor Italo Giglioli, director of the “Regia stazione chimico-agraria sperimentale” (Royal Experimental Chemistry and Agriculture Station) in Rome, published a book both presenting the many problems that limited the modernisation of Italian agriculture and indicating the solutions to be adopted (Giglioli 1903). In the specific case of oenology, the way forward was to improve technical knowledge and industrial practices, as the achievements of enterprising Californian oenologists demonstrated. Italy could become relevant in the international wine trade by opening up to technological progress only. Some good results began to arrive, as demonstrated by the growing trend of exports of bottled wine, the highest quality and destined for a more demanding clientele: however, these results only concerned the best wineries and not the most of winemakers.

At the beginning of the twentieth century, new important steps to improve the winemaking were related to the adoption of the new machines and the modernisation of cellars. The change, as usual, involved a minority of producers: a large part of the peasants continued to follow old procedures and made poor-quality wine, sometimes barely drinkable. The State tried to help the renewal of the sector by the foundation of new schools to train technicians in oenology. The oenology became fashionable through the proliferation of popular publications and newspapers where novelties arriving from abroad were published and discussed by readers, that is the new generation of Italian agronomists. They represented the élites in favour of innovations and, with their example, they had to spread the modern ampelographic knowledge in the countryside and allowed the real improvement of the productive system and, by this way, the increasing of wines which could be appreciated outside the local taverns.

2.3 International Promotion: Oenotechnical Agencies and Cellar-Stores

From the 1870s to the First World War, central, provincial, and municipal Italian public administrations promoted the modernisation of Italian viticulture and winemaking, as well as the internationalisation of Italian wines. They played a fundamental role for: (a) the transmission and transfer of knowledge and information about the production system (winegrowing and winemaking); (b) the control of the final product (to improve its quality and avoid alterations); (c) the distribution of Italian wines on the international market, giving wine producers all the information about customers' tastes and the potential demand concerning Italian wines (Stringher 1901; Musella 1992; Zaninelli 1990). During the analysed period, the initiatives of the State, the Provinces and the Municipalities in favour of the creation and the diffusion of a new national wine which was adapted to the international market were numerous: they helped the creation of a national oenology which was able to face an increasingly globalised wine market.

These public institutions in particular financed the creation of: agrarian councils (*comizi agrari*) which supported studies aimed at the development of the agriculture (obviously including the viticulture); agricultural schools training new qualified farmers as well as new winemakers; experimental agrarian stations (*stazioni agrarie sperimentali*) conducting scientific experiments on the primary sector; and itinerant agricultural chairs (*cattedre ambulanti di agricoltura*) whose agronomists explained in the rural villages the new efficient production systems, including new methods of viticulture and winemaking. This policy was accompanied by the dissemination of agricultural almanacs and agricultural journals: the aim was evidently the improvement of the production and productivity of the Italian primary sector (Fumi 1999; Pazzagli 2008; Vaquero Piñeiro 2015; Locatelli and Tedeschi 2018).

These initiatives guaranteed a considerable help also for the oenological sector, but they were not able to recover the backwardness existing in comparison with other countries and in particular France. Specialised schools for the vocational training of future winemakers were created: their courses allowed new oenological technicians to easily find a job in the vineyard farms (as well as in the agronomic institutions previously indicated).

The public contribution to the improvement of the Italian wine sector also involved the promulgation of laws favouring the birth of social wineries (*cantine sociali*). The first ones were cooperative wineries associating wine producers and assisting them during the winemaking (to improve the quality of the final product) and the marketing (to reduce distribution costs related to the sale). The second ones were organised by the Italian government: they rented the machines for the winemaking process (pumps, presses, boilers, etc.) and this reduced the production costs of the grape must (Maffi et al. 2020).

Finally, the Italian State financed the creation of free warehouses (*magazzini* or *depositi franchi* exporting Italian wine at reduced customs tariffs in some European countries) and especially oenotechnical stations (*stazioni enotecniche*) which were installed in the main European cities with the aim to promote the export of Italian wines (*Depositi governativi* 1883). Their managers gave Italian wines a precise identity on the market, even if different wines were produced according to the diverse terroirs of production. The Italian oenotechnical stations in America were important for both Italian and local viticulture: the presence of Italian emigrants in fact favoured the sale of Italian wines, but it also contributed to the progressive improvement of local wine production. In the case of the Californian oenological sector, emigrants brought their know-how and the exchanges of information favoured the arrival of new techniques and tools in the Italian terroirs; in the case of Argentina, migrants developed the local oenological sector and Italian schools for winemakers gradually became a model and a real point of reference. The activity of these oenotechnical stations clearly showed that the role of the Italian State as an agent of innovation in the nineteenth century was not limited to investments seeking to improve the winemaking.

Before the public intervention, the novelty of the wine sector was the establishment of the so-called winemaking societies. Born in the years before the unification of Italy, following the main relevant landowners' will to encourage the production of quality wines, these societies did not really modernise the Italian wine sector and so they failed to respond to international competition and the new strong protectionist customs policies. Nevertheless, they still had the merit of encouraging the improvement of winemaking techniques.

At the end of the 1880s, the negative consequences of the "tariff war" against France (the main foreign market of Italian blended wines) obliged public institutions to create a close interaction between private and public initiatives. This collaboration was not limited to the transmission of knowledge and information concerning new modern production systems (vineyard crops and winemaking) and the control of the final product (to improve its quality and avoid adulteration), but it also promoted the establishment of new institutions and the implementation, in Italy, of modern entrepreneurship and business models oriented towards international wine production and marketing.

In 1884, the General Society of Italian Winemakers was born in Rome. It was promoted by the Minister of Agriculture Domenico Berti (Atti convegno 1884). The aim was to identify appropriate measures to improve the wine sector which needed a real process of tools and techniques' renewal. In the Italian parliament lacked a coalition defending the interests of the wine sector. The cereal groups carried the most weight. The association reached 1800 members, thus demonstrating the successes of combining public policy, private interests, and production expertise. In 1885, Professor Giovanni Battista Cerletti, the founder of the school of winemaking located in Conegliano Veneto, was appointed Secretary-General of the Association. This association remained active until 1897 (when it joined the Society of Italian Farmers) and always supported the implementation of new laws in favour of the Italian wines. The association edited many books and contributed to the foundation of social and experimental cellars, as well as the development abroad of "oenotechnical agencies" and cellar-stores (Rogari 1994).

The Government intervention supported the implementation of a structure devoted to produce and promote quality wines to ensure the access of Italian wines in foreign markets. Some "Stazioni Enologiche" (literally "oenotechnical stations/agencies") were opened in major European and American cities providing a substantial contribution to the promotion: Lucerne in 1884 was the first, followed by London and Munich, in Europe, and by New York and Buenos Aires in the Americas (Vaquero Piñeiro and Maffi 2018; Vaquero Piñeiro 2019). Led by qualified technicians, these agencies had to collect information on the conditions and characteristics of the different markets and provide Italian manufacturers with advice and instructions to access those markets successfully. The constant quality checks run to avoid any counterfeits, thereby guaranteeing the quality and the origin of wines to importers was another important function they had. The "oenotechnical agencies" were excluded from any trade negotiations to ensure their impartiality. Their role was to create strategic linkages between shopkeepers and importers, to collect their complaints and suggestions, and to facilitate the introduction of Italian wines in foreign markets. By visiting restaurants and hotels, they provided additional information about customers' tastes, sale prices, and the effects of advertising on consumption. Data were sent to the Ministry of Agriculture and thereby to the producers: so, they could take action to remedy the shortcomings and implement the best practices. The agencies also had to buy samples of the best-selling non-Italian wines to analyse them: the aim was evidently to understand the reasons for their commercial success.

The directors of the "oenotechnical agencies" played a key role in the creation of the image of "the good Italian wine" abroad: they selected the best Italian wines for the export, in terms of quality and oenological peculiarities. Through the "oenotechnical agencies", especially those installed in the Americas, the Italian Government facilitated technicians training and innovation.

To compete with foreign producers (in particular the French), the public policy supported the implementation of specific schools for wine technicians, winemakers, and oenologists. The courses allowed students to learn the latest news related to ampelography studies and modern winemaking techniques. A further contribution was made by enacting laws in favour of collective cellars and warehouses reserved for agricultural machinery. Collective cellars were cooperatives that brought together wine producers and helped them in the winemaking and marketing process (reducing their costs). Warehouses rented what was necessary for the winemaking process (e.g. modern pumps, presses, and boilers), reducing the grape must cost. In addition, the Government supported the participation in international exhibitions related to the wine sector.

The "oenotechnical stations" tried to take advantage of the market demand resulting from the massive presence of Italian migrants abroad: they offered Italian wine producers a detailed picture of what the most innovative competitors produced. Thus, the head of the New York agency pointed out that the fast growth of winemaking in California was linked to the use of high-tech machines and production processes. It was important for Italian winemakers to be informed about this sharp increase as they had to learn new marketing techniques and consider the relevance of advertising as quickly as possible.

Along with the "oenotechnical agencies", in several European cities, such as Lucerne, Frankfurt, Munich, Berlin, Paris, London, and Hamburg wine cellar-stores opened. These shops benefited from reduced customs tariffs and were contracted to a local company that assumed the obligation to sell only Italian wine. The wine sold in bottle could also be cut to get a product suitable for the demand; in this case, the Italian firms agreed to send bulk wine by train using the tanker wagons. The wines had to be analysed by the wine technician who, in the end, was responsible for the marketing of wines from Italy. Deposit dealers had to focus solely on the wholesale, leaving retail to local sellers. In the concession documents, the Italian Government required the owners of the cellar-stores to carry out an intense advertising action, using any available tool.

The oenological stations in the United States and Latin America sought to take advantage of the presence of Italian emigrants who guaranteed a sure demand and also helped producers to distribute Italian wines in other migrant communities. In addition, they observed the practices of local producers and sent to Italy relevant information for the renewal of Italian viticulture. Furthermore, they helped Italian peasants (in particular the viticulturists) who had to emigrate: they in fact organised training courses in viticulture and winemaking to improve their ampelographic knowledge. So the most important oenotechnical stations were located in New York and Buenos Aires, that is in great countries with large wine markets and a lot of Italian migrants.

The wine station in New York played a key role in defining the characteristics of Italian wine and gave Italian public authorities and winemakers the opportunity, through regular information, to better know the wine sector in a country where the principles of modern industry were applied to winemaking. For the Italian viticulture of the late nineteenth century, which remained attached to the artisanal production, there was also another real challenge: it was in fact necessary to help the Italian migrants who could potentially become competitors but who also favoured the diffusion and consume of the best Italian wines.

The New York Wine Station was entrusted to Guido Rossati in 1895. This former representative of the Italian Government, having the charge of the promotion of Italian wines in London, had for mission to get the characteristics of the Italian wines in England better known by consumers. In the United States, he devoted himself to publishing articles on Italian wines in the local press, and he was also the author of a vast study on grape growing and wine production in the United States. In this volume, the result of an expedition across the country, he provided relevant information on the viticulture of each State of the United States, emphasising the environmental characteristics and the quality of the railway and maritime infrastructures which he considered as fundamental to increase the export. He dedicated a particular attention to the development of viticulture in the California districts of Sacramento, Napa, Sonoma, Santa Clara, Livermore, San Joaquin, and Los Angeles. He also verified the entrepreneurial activities carried out by Italian migrants, such as the Italian-Swiss colony of Asti: these latter gave an essential contribution to the development of the US wine industry and the Italian-Swiss colony was in particular responsible for the introduction of Italian red wines in California, such as Barbera, Nebbiolo, Muscat, Bonarda, and Refosco which were sold under the name Crabb's Black Burgundy (Rossati 1900).

Rossati's works show that the role played by the Italian State in favour of the process of internationalisation of the Italian wine industry was relevant (even if sometimes underestimated). Rossati indicated that environmental conditions made California an ideal region for the production of large quantities of ordinary wine: he also underlined the superior technological capability of American wine producers. Their mechanical industry designed and built new machines carrying large quantities of grapes: at the same time, the automation of some procedures allowed a considerable labour saving and reduced the cost of the winemaking. According to Rossati's opinion, Italian viticulture had to follow the example of Californian winemakers: so Italian ones needed to obtain new information through magazines and specialised publications or, better yet, they had to organise study trips to observe the new Californian facilities that make wine production easier and more efficient, in terms of saving time and manufacturing costs. Rossati's book illustrated important photographic material showing the machines and buildings of the Californian wineries: he thought that these latter represented a model to follow for the Italian producers and his photos showed the state-of-the-art of Californian winemaking, with a greater emphasis on storage operations, on the rational disposal of vineyards and on new grape must processing facilities.

For the most enterprising winemakers, news arriving from the United States were a point of reference and represented a constant stimulus for innovation. Rossati thought that the press was important for the success of the Californian viticulture, both for the feature articles and for the announcements allowing a wide dissemination of information. The publications dealing with the problems of the wine industry were distinguished by important draws, documented articles as well as the originality and modernity of their pages (fonts used, images presented etc.). The factor that Guido Rossati attributed with the success of viticulture in California was the role of the press. The publications that addressed the problems of the wine industry were distinguished by their large print runs, the originality of their features and documented articles, and an abundance of advertisements. Special mention was reserved for the *Pacific Wine and Spirit Review* and *California Hotel and Wine*, both published in San Francisco, and the *American Wine Press* from New York. The same considerations also applied to the Italian language newspapers printed in San Francisco, in particular *Italia*, edited by Ettore Patrizi and G. Almagià, *Voce del popolo*, and *Elvezia*, which was aimed at the large Swiss colony from Ticino that had settled on the Pacific coast. These were newspapers intended for immigrants, which had a clear function as a point of reference, and for the transmission of information beyond the barriers imposed by the use of English. Thus for Italian immigrants, California offered the conditions for practising a more advanced viticulture, which also demonstrated the road ahead for Italy. In conclusion, Guido Rossati's book constituted a clear example of how Italian viticulture needed to draw lessons from the situation in California, as an impetus and stimulus for change.

Rossati underlined that also in Italy, the oenological sector disposed of newspapers and specialised magazines: it was necessary to renew them by copying the Californian ones. The same suggestions were given to newspapers which were printed in San Francisco in Italian language and were addressed to migrants: they were more relevant because they allowed migrants to receive information overcoming the linguistic barriers (i.e. the use of English in main important American magazines and journals concerning the winegrowing and winemaking). Rossati considered that California offered to Italian migrants all the best conditions to practise a more innovative viticulture. Their good results obviously increased the competition in the US wine market which, in the early twentieth century, continued to grow up and consumers demanded to Italian winemakers (producing in Italy or in California) to increase the quality of their products (Vaquero Piñeiro and Maffi 2018).

Finally, for the new oenologists who graduated from new Italian agricultural schools, the growing American viticulture also offered excellent professional possibilities: it was obviously necessary to study the English language which was not learned at the oenological school. They had strong plant breeding skills, but also a limited knowledge about the machinery turning traditional wineries into modern and efficient ones. Even the simple use of steam engines in fact represented an absolute innovation for most of the Italian oenologists: they sometimes accepted a job in a mechanical factory where they can improve, but most of them found a job in the best Californian wineries and helped these latter to improve the quality of their wine.

An important role was made by the first enotechnical station in Buenos Aires which went into operation in 1888. The first director, Pompeo Trentin, a graduate from the oenology school of Conegliano, was the author of several volumes on wine trade (Trentin 1895) and carried out widespread activity in dissemination as well as study in order to learn about the conditions of viticulture in Argentina. The report written during his trip to Argentina and Chile was very important because the text included drawings full of information and first-hand news on the history and conditions of the vines and wine in two areas of the Andes at the end of the nineteenth century. Pompeo Trentin was responsible for sending systematic, detailed reports to the Italian authorities regarding prices, currency, and workers as well as the trade journal, legislation and strategies to be followed to increase the visibility of Italian wines. His abundant observations were regularly included in the Agricultural News Bulletin and in other publications of the Ministry of Agriculture. Today, these are a first-hand source for studying the consequences of the 1890 financial crisis not only in the wine sector but also in the whole Argentine economy. In 1896, Trentin promoted the celebration of the first Italian wine exhibition in Buenos Aires.

According to Trentin, the Argentine market presented very favourable characteristics not only for Italian cut wines. The wines best suited to Argentina's demand, however, had to meet a series of requirements: low alcohol content, ruby colour, and fresh arome and excluding excessively sour wine. Therefore, to produce wines that had the characteristics suitable for the Argentine market, it was necessary to learn how to make "blends" by combining different types of wines until the right one was found. The work of the director of the enotechnical station shows how the globalisation of trade, especially in the agri-food sector, required making products that were suited to the tastes and expectations of international customers (Ceccarelli et al. 2013). Today there is an overall interest in typical products and territorial identities connected to food. However, looking at the history of Italian wine in Argentina, it is clear that at the end of the nineteenth century the main problem that needed to be overcome was that of eliminating specific local characteristics in order to create uniform products that were destined for the international market (Magagnoli 2015).

The director of the enotechnical station insisted that exports could not depend solely on the ethnic and sentimental factors of immigrants. Reference to the food of the "homeland" could predominate among the working class (Cinotto 2013), but at a higher social level these constraints were slackened and it became essential to offer products safe from the standpoint of people's health, while also being pleasing and captivating. Therefore to explain the trends of the statistics behind the wine trade during the first phase of internationalisation between the late nineteenth and early twentieth centuries, it is necessary to consider the action of many players, such as the directors of the enotechnical stations, which in addition to carefully following the markets, gave Italian wine a precise and recognisable identity. It is no coincidence that Pompeo Trentin spoke to numerous wine producers before taking charge of the Buenos Aires enotechnical station. He collected information on the different types of wines produced in the Italian peninsula and this knowledge proved essential once he settled in the Argentine capital. From this point of view, Pompeo Trentin and his successor Tommaso Chiaramonte symbolise the changes that occurred worldwide in the wine sector. Both technicians, appointed by the government on the basis of their skills, had the task of fostering the internationalisation of Italian wines by encouraging a more modern and competitive way of wine marketing. Even these aspects should not be overlooked when talking about including wine in the globalisation of trade. The Italian oenological station in Buenos Aires was in operation until the early 1920s.

While Pompeo Trentin was encouraging the adoption of trade strategies suitable for transoceanic sales, he also witnessed the role being played by Italians in the birth of the Argentine wine industry (Scarzanella 1986, 91–109). The contribution of immigrants to the development of the Argentine industrial system is one of the distinguishing features of the history of the national agri-food sector (Barbero 2008). Regarding the oenological industry, which had no barriers and in a climate of great entrepreneurial freedom, Italian immigrants found the ideal conditions to put into practice a solid background of knowledge and experience (Girini 2014). In 1896, the Escuela Nacional de Vitivinicoltura began to work in

Mendoza following the example of those working in Italy (Rodríguez Vázquez 2012, 539–558). In the context of the circulation of knowledge and skills coming from Europe, Argentina is a concrete example of the formation of the world wine industry. In 1903, 27 of the 34 wine-producing companies operating in Mendoza were in some way attributable to Italians (Barrio 2008, 81–111). At the end of the nineteenth century, the Argentine winemaking sector was made up of two realities: on the one hand a national industrial infrastructure that was growing rapidly in terms of the production of wine in large quantities, and on the other, a growing market which required Italian companies to adopt strategies that abandoned the most common wines and moved in favour of mid- to high-end wines.

Success was linked to the virtuous collaboration between national institutions, wine associations, and individual entrepreneurs. The information realised by national institutions encouraged producers to modify products to be sent to the internal as well as to the international market. The "oenotechnical agencies" and the cellar-stores allowed public policy to indicate the way to follow and determined the characteristics of the Italian wine trade in the world. The confidence of some entrepreneurs in the suggestions received and the consequent investments to improve production techniques led to an improvement in the quality of Italian wine. This quality increase enabled best Italian wine producers to export wine to all the major international markets.

In the context of the globalisation of the wine market, the renewal of viniculture and winemaking methods was crucial to make Italian wines competitive. Quality was essential, but the diversification of the types of wines was also important. The link between Italian politics and the winemaking sector was, therefore, not limited to the presence of three important Presidents of the Council who were, at different times, important wine producers (Camillo Benso of Cavour, Bettino Ricasoli and Agostino Depretis), but also depended on élites' awareness that the Italian economic development could not ignore the primary sector, including the viticulture. The increase in wine sales on international markets supported both the trade balance and the national prestige. The choices made by public institutions allowed entrepreneurs to invest in an international perspective, thanks to the opportunity of creating relationships and getting information from the wine market in the most important and evolved towns in the world: great opportunities were offered to the best Italian wine producers.

2.4 Legislative Intervention: Fraud, Health, Typical Wines Versus Regional Appellations

Despite the intense agronomic propaganda behind it, Italian wine enjoyed little fame on the 1862 World's Fair in London. Leaving aside liqueur wines such as Marsala or Vermouth, which were very appreciated by foreign markets, the London event was a real watershed for the Italian wines. However, although statistics showed a very limited increase, the first decades of the 1860s and 1870s were very important: some relevant initiatives in fact matured to encourage a different approach to vine cultivation and wine production. Another important innovation was the creation of a ministerial commission to taste and evaluate the quality of wines to be sent to the Vienna Exhibition of 1873.

The commission had to correct the mistakes made at the London and Paris Exhibitions, where Italian wine producers had participated without any preparation or knowledge of the subject. Most of them in fact fell into the error of "confusing small, local or even regional exhibitions with large international exhibitions": they thought that the difference was essentially the size of the venue and the number and variety of products on display. They did not realise that, although they had good-quality wines for the Italian market, in the international exhibitions their wines were getting mediocre ratings. This also created a negative prejudice concerning Italian wines damaging the few products having some chance of standing out.

Another serious defect highlighted by the ministerial commission was the extreme variety of names given to Italian wines related to the traditional shortcomings in terms of trade. Some wines had fantasy names, others indicated the areas of origin or the grape varieties used, and in a rather innovative way, the ministerial commission addressed for the first time issues concerning the relationship between the wine and the land where vines were cultivated. Even if the concept of terroir did not exist, some aspects were present in the background, and they would later take shape during the twentieth century. The commission in fact wrote: "to be a wine defined as typical, apart from having special and constant qualities which make it easy to distinguish from others, it must be made in large proportions; this is what many people ignore and which forms the weakest side of our wine-making industry. Each producer believes he has the right to impose on his wine the name that his whim dictates, and often, because of the mania to make many wines in small proportions, the names multiply, generating confusion without purpose. Sometimes the wine has the name of the small farm where the grapes were harvested; in other cases the names are taken from history, mythology and even from the producer's family… Some districts have already acquired fame for their vines and their wines, they are sufficiently important to be geographically known, so that the name given to the wine is equivalent to a brand that accredits it and almost indicates its quality, since when the origin of a wine is known, its distinctive characteristics are also known".

The "contrade" and type wines mentioned by the commission were Barolo, Nebbiolo, Grignolino, Barbera, Bonarda, Caluso, Chianti, Montepulciano, Pomino, Lambrusco di Modena, Valpolicella di Mantova, Vernaccia di Sardegna, Moscato and Marsala di Sicilia, Lacrima di Calabria, and Capri. French authorities had already understood the economic importance of protecting the designation of origin as it offered the buyer the certainty of the quality and origin of the product. This position was strengthened by the signing in Madrid in 1891 of the international agreement on the rules concerning the registration of trademarks for the productive and commercial companies: Italian government did not adhere because of the use of the name cognac having a great diffusion and commonplace in Italy. The denomination of origin had clear economic implications, but the government's choice avoided that in Italy had concrete consequences.

New laws were promulgated at the end of the nineteenth century: they wanted to regulate the characteristics of the wines to avoid negative consequences for people's health (Lonni 1998). The aim was to reduce fraud and manipulation, even though artificial wines resulting from formulas and mixtures of various natural and chemical substances and ingredients were widespread. The preparations allowing everyone to make their own wine without using grapes received a lot of publicity: manuals explained how to compose the right mixtures in order to produce low price drinks similar to a good French red or German white wine, without neglecting the possibility of achieving the even more coveted champagne. Another popular gimmick was to use dried grapes to make sugary liquids which were then mixed with a southern wine and sold as genuine wines. Artificial wines constituted a recurrent practice and they were consumed by less purchasing power people (Vecchi 2017).

While the manufacture of artificial wines gradually declined as wine culture and sensitivity to people's health increased, wine frauds continued because they were the producers' answer to high taxes concerning the introduction of the wine into the "wall of towns" (Lunardoni 1904). The duties paid at the city gates increased the price of wine discouraging its consumption among the lower classes. At the same time, new duties prompted wine producers and tavern owners to mix wine with water. The watering down of the wine meant that the wines on sale lacked credibility. In short, fiscal policy (Moricola 1998) favoured neither an increase in wine consumption among the working classes nor the arrival on market of genuine wines. In the background there were always negative judgements on the excessive consumption of alcoholic beverages, considered bad for morals and civil coexistence. Public authorities followed contradictory goals: they needed to increase wine consumption because wine-growing and winemaking guaranteed important earnings for the primary sector and increased the tax revenues: at the same time, they attempted to introduce limits to wine consumption (Hames 2012).

Public authorities also had the problem to reduce the risk of Italian wines being counterfeited in foreign markets. Even if there is no comparison with the current situation (Italian sounding food and drink have a global turnover of around 100 billion euros), at the end of the nineteenth century "real vermouth" from Turin, but actually made in Portugal or Peru, was circulating in North and South America, as it also happened at a false "Moscato d'Asti" made by American wineries skilled in fermentation techniques giving the drink a pleasantly sweet, sparkling flavour. Another market specialising in bottling false Italian vermouth and sparkling wine was the German port of Hamburg: the labels requested by buyers of bottles were normally affixed.

The first legislative intervention concerning wine was the 1890 law for the hygienic supervision of foodstuffs, beverages, and household objects: it prohibited the manipulation and addition to wines of substances dangerous to people's health. In 1900, the first law on wine fraud was promulgated: it helped to mark a radical change in the wine consumption. Besides, following the Madrid Convention of 1891 on trademarks, the debate on the protection of typical wines arrived in Italy: once again it was the Chianti area that led the change (Mocarelli 2013). In 1902, an association was set up to promote the Chianti trade and in 1903 a "Sindacato enologico cooperativo del Chianti", that is a Cooperative Winery associating winemakers of the Chianti area, came into being. Other legislative interventions on wine fraud took place in 1904 and then in 1909 when the law on distilleries came into force (Benatti and Castelletti 2022). Italy followed other countries engaged in introducing rules to manage the wine sector having wide-ranging commercial and health implications for people

(Da Silva Lopes et al. 2020; Stanziani 2004, 2005). There was a range of frauds, falsifications, imitations and adulterations in the beverage sector, and in particular with wine, which made up a panorama demanding the adoption of appropriate measures. Other laws against frauds were promulgated in 1916, in 1917, and moreover in 1924 when, on Senator Arturo Marescalchi's initiative, the thorny question of a law on the definition and protection of typical wines was tackled (Marescalchi 1926). To avoid the hurdle of having to deal with difficult delimitations of the production areas, the law was oriented towards the concept of protection through the establishment of special consortia: they had the right to establish specific rules to ensure the best quality of each type and create a strict relationship with the name (Kovatz 2013). So, the Royal decree law no. 497 of 7 March 1924 gave producers of a particular typical wine the right to form a consortium to protect the name of their product and in general to achieve the aims set out in the new law as well as to adopt their own brand or distinctive sign to distinguish the products of the consortium members. One of the first consortia to be founded was that of Chianti, but it was not relevant because of the absence of a precise delimitation of the production areas. De facto, the law was not fully applied: there was no precise definition of "typical wine" and no clear rules existing to establish the borders of the production areas and the related use of the brand name.

These aspects explained why in the following years a heated controversy erupted over the use of the name "Chianti". Some producers wanted to reduce the denomination of origin from seven designations to just one: this simplification was claimed because Fascist regime was contrary to the "anti-productive jealousies of the factions". However, there were constant reports about falsification to the detriment of "Chianti" producers. The clash began as soon as the 1924 law on consortia and typical wines was approved (Benatti and Castelletti 2022). The first law concerning protection of the designation of origin in fact authorised the foundation of a consortium for the defence of Chianti. However, this law did not define the characteristics of the wine or delimit its production area. As a result, anyone in Italy could bottle and sell a "Chianti" wine (or any other wine) without taking into account either the type or the origin of the grapes used (Stanziani 2005; Garcia-Parpet 2009). Among the reasons that prevented a law to regulate appellations of origin, the political climate prevailing in the country had to be taken into account. On the one hand, there were producers from the southern regions who were specialised in selling grapes, and on the other hand, Tuscan Chianti producers who were interested in the external market (Mocarelli 2013; Mocarelli and Vaquero Piñeiro 2017). This made it impossible to reach an agreement compatible with the needs of both parties. In this context, wines of mediocre quality prevailed. In 1925 the Consortium for the Defence of Typical Chianti Wine registered its trademark with the "black rooster" and in 1927 the Consortium of Chianti Wine did the same, using a "putto" as its symbol. A massive demonstration of more than a thousand people was organised in 1935 in Siena in favour of the creation of a single brand and reference territory overcoming the division between the Chianti "Gallo" and "Putto" consortia (Corriere Vinicolo, n. 9, 2 March 1935). This division has dragged on for a long time even after the distinction between "Chianti Classico DOCG" and "Chianti DOCG", the former produced in the historic area of the same name, the latter in the wider territory between the provinces of Florence, Prato, and Pisa. A long battle explaining the late arrival of the DOCG for Chianti only in 1985 and the DOCG for Chianti Classico in 1996. The controversial case of Chianti clearly shows the difficulties existing in Italy in introducing regulations to govern the wine sector, fragmented between a network of territorial and commercial interests.

Fascist regime wanted to regulate every aspect of the national economy: this led to the drafting of many laws on wine sector and related matters. A new law in 1925 once again concerned the repression of frauds in wine and vinegar productions. It was also made a precise definition of wine as "the product of alcoholic fermentation of the must of fresh or slightly dried grapes in the presence or absence of marc". It was a clear attempt to introduce clarification in the alcoholic beverages sector, full of nuances favouring bad producers. The trade and sale of non-genuine wines was prohibited and this prohibition also included wines with an alcohol content of less than 9 degrees for red wines and 8 degrees for white wines. However, the trade regarding the so-called vinello, that is, wine with a light alcohol content, was authorised because it was a very popular drink (Marescalchi 1926).

In 1925, with the proclamation of the "battaglia del trano" (literally "Battle of Wheat") the Fascist regime, which had initially maintained a liberalist attitude in the rural sector, brought about a radical change in the national economic orientation: new regulations strongly protected the Italian productions of wheat provoking in the foreign markets the increase of tariffs on Italian food and drinks (Marescalchi 1925). Difficulties also increased due to the general fall in agricultural prices after 1926 (Fabiani

2015). No aids were provided by public institutions: so, many small farms were forced to close, crushed by the existing debts and the falling sales prices. The attempt to redress the balance of payments was based on the improvement of the cereal production, without taking into account the need for alternative production guaranteeing a positive contribution to the foreign debt position. The Italian agriculture was subordinated to the principle of strengthening and full use of national resources; in the name of food self-sufficiency and the reduction of imports, the state arrogated to itself an absolute decision-making capacity and almost completely eliminated the rules of the free market. The main effect was the transformation of rural enterprises into simple passive executors without any margin of autonomy.

Producers, both large and small, had to respect the fascist regime indications, which, with insistence and a profusion of propaganda, concentrated all its interest on increasing wheat production. Everything else took a back seat, including attention to other crops, such as vines, which, compared to wheat, required greater capital investment, long lead times, functioning markets, and a considerable element of risk. In this political context, the interest of agricultural newspapers for grape cultivation and wine quality almost disappeared. By the end of the 1920s, the "patriotism" of every single farmer in Fascist Italy was only measured in quintals of wheat harvested.

In the early 1930s, the difficult situation following the economic crisis of 1929 obliged the Fascist regime to decide on new measures to help the wine sector whose dimensions, even if they were reduced by the policies followed in the previous years, remained relevant. Vines occupied about 4 million hectares of land, even if it was partly associated or alternated with other crops. Farmers annually invested in vineyards and cellars about 32 billion lire employing an impressive workforce, that 450 million working days every year. The first measures concerned the reduction of transport costs for southern wines, the exemption from the wine tax, the foundation of cooperative social cellars and "enopolies". A new law for the defence of typical wines was also prepared. Because of a negative economic trend and the related falling prices, closed markets and strong reduction in purchasing power, no suggestions regarded the search for better quality wines. Producers simply tried to preserve the positions acquired, without dispersing the heritage of knowledge and sensitivity arriving from the past, that is the increasing of quality of the beginning of the 1920s. Public authorities had to admit that most Italian wines had low quality and, moreover, that the special wines, although appreciated for their high oenological characteristics, represented a small part of the total production and had a low economic and commercial importance.

To combat the crisis of the wine sector, weighed down by falling prices, full cellars and the progressive debasement of the production, the main solution arrived by new regulations encouraging the "corporate management", as happened to for the whole national economy after the creation of the new public guilds (*Annuario vinicolo d'Italia*, 1941–1942). The aim of laws of 29 December 1930 no. 1737 and 6 January 1931 no. 22 was in fact to create winegrowers and winemakers' associations and to set up cooperative cellars and enopolies in order to make rational use of the by-products of winemaking (Benatti and Castelletti 2022). In order to reduce the negative effects of the overproduction of wine and to have more wine not fit for direct consumption, a new special tax treatment was granted for wine alcohol used in the preparation of cognac and vinegar and for the alcohol obtained from the distillation of "vinello". Furthermore, in application of Law No 987 of 18 June 1931 on plant protection, 62 provincial consortia were set up. They provided training for workers, assistance to winegrowers, and so on: they also had the power to set up consortia enopolies for production areas in order to collect wines from producers and to unify main relevant characteristics of each individual terroir. The law contained other measures to protect wine-growing, including provisions on production discipline and consumer protection: in particular a ban on selling vinegars that were not made from wine and musts that were not really made from grapes. At the same time, in order to encourage the consumption of fresh fruit, the "Grape Festival" was proclaimed.

In the early 1930s, the "Office international de la vigne et du vin" identified the production of typical wines as one of the strategies to be followed in order to revive the world wine sector, which was suffering from the strong negative economic trends. Observing the suggestions coming from the office and the new legislations that France and Spain had adopted to defend their wine (Planas 2016), in Italy new rules were promulgated (Royal Decree of 11 January 1930 n.62) for the defence of typical wines and to promote the development and the improvement of their production as well as their diffusion on national and foreign markets. It established a new definition of typical wines, that is those genuine products from certain localities whose grape variety and vinification method have clearly definable particular and constant characteristics. Wine producers and merchants were encouraged to form consortia for the protection of their wines. While the parliament was recommending to be left free of all constraints of the production and sale of Vermouth, in accordance with the major producers' wishes, the Marsala was excluded from this approach because it had to be protected by law. The production areas of best typical wines such as Marsala, Orvieto, Moscato d'Asti, Asti spumante, Soave, Castelli Romeni, Chianti, Barolo, Barbaresco, Capri, Lambrusco, Sangiovese, Albana, and Freisa were delimited (D'Agata 2019). A specific action was taken to defend the zibibbo grape and Moscato di Pantelleria. The Chianti territory, the best-known Italian wine, was divided into seven production zones, demonstrating the enormous difficulty of reaching a synthesis capable of overcoming the strong roots of territorial particularities: Chianti Classico, Montalbano, Rufina, Colli Fiorentini, Colli Senesi, Colli Aretini, and Colli Pisani.

The legislation on wines was then completed with Royal Decree No 1443 of 26 October 1933, which extended the national mark for fruit and vegetables to export (established by Law No 1272 of 23 June 1927) to wines, in order to give buyers of foreign countries a guarantee of the authenticity of purchased wines. In the context of a specific law on the defence and definition of typical wines (Mistry and Valente 2021; Rossi and Cortassa 2021), the consortia were assigned the task of defining typical wines through the delimitation of production territories (Image 2.2).

The new legislation gave rise to a number of new problems: it was necessary another law in 1937 on "Provisions for viticulture and wine production" including the "Discipline of the production and trade of fine wines of specified origin". In the context of the fascist autarkic regime (after the Italian invasion of Ethiopia and the related Leagues of Nations' economic sanctions) this law represented a further step towards the full regulation of the wine sector: fascist regime aimed at reconstituting viticulture in terms of improving the quality of wine and, moreover, table grape production. All peasants wishing to plant or reconstruct previous vineyards in fact had to obtain authorisation from the local consortium. Fascist regime wanted to have the total management of wine production and trade to increasing the production and packaging of fine and superior quality wines in order to eliminate foreign wines from the Italian market and to increase wine exports. Other objectives of the law included the elimination of defective and low-grade wines from the domestic market as well as the fight against adulteration and fraud and the defence of the guarantee marks for fine wines. These measures also allowed to defend prices by the improvement of quality and, moreover, the production of the quantity that actually markets needed.

Image 2.2 Students doing wine tasting exercises (1930s). (Source: Archivio Storico Indire, Fondo Fotografico)

Through the law of 1937, Fascist policy for the wine sector wanted to confirm the economic and cultural dimension of Italian wine production. It aimed to regulate the market for grapes and wine in order to ensure that producers received fair remuneration and were not subject to the frightening fluctuations caused by the changing production trend. Important in terms of balance remained the use of wine surpluses in distillation for the production of alcohol. Both the oenopolies and the cooperative wine cellars played an important role in calming the market: they relieved producers of the worry about the placement of grapes by staggering the sale of wine over the course of the wine year.

However, the law of 1937 did not fully come into force, but it created numerous problems because the 1930 regulations were repealed, creating a legislative vacuum. So, the wine sector reverted to its original position and Italy remained without a legislation protecting typical fine wines. This opened up a wide-ranging debate that led to countless proposals. The concept of "typical wine" had to be overcome and many doubts were also advanced about the definition of "fine wine", which was considered too ambiguous and not linked to the terroir. So, it emerged the idea of giving priority to the territorial origin of wine, that is the link with the place, culture, and history, matured. However, these themes were actually taken up in the 1950s only.

Another problem in fact affected the debate concerning the wine sector in the 1930s, that is the use of pesticides. Farmers had the bad habit of spraying their vines with a solution of copper sulphate: a high quantity of copper sulphate or vitriol was used. The doses were excessive, it was a real economic waste, and the fascist regime wanted to modify the situation by teaching how to prepare the mixture in a more balanced way. The problem of copper sulphate intersected with the autarkic policy, which made it increasingly difficult to obtain sulphur and copper. Against this backdrop, journals concerning winegrowing published advertisements for less expensive products which were made by Italian factories. Other substitutes had to be found for materials for the viticulture which were in shortage on the Italian, such as galvanised wire, supports in reinforced cement, or even dynamite, which was used to break up particularly stony ground.

On 20 September 1938, a new decree came into force: it obliged to report the quantity of wine produced, requiring that 10 per cent of the production be set aside to safeguard farmers' interests. The aim was to regulate the price fluctuations on the wine market. Besides, new regulations limited the production where winemakers did not realise fine wines or encouraged improvements in quality by promoting the establishment of cooperative wineries where winemaking disciplines were applied with "prudence and seriousness". In 1939, a technical committee was set up to examine the documentation produced for the declaration of fine wines having a specific origin. New limits concerning the wine tried to reduce the wine on the market, in order to keep prices stable. At the same time, the use of marc was encouraged: in the logic of making full use of national resources, it had to be used for the production of distillates, alcohol, fats, colouring materials, and cream. On the eve of the Second World War, using the slogan "no price increases for any reason and in any form", the corporate organisations intensified controls and created local commissions carrying out continuous checks to ensure that all consumers, sellers and producers diligently complied with the price regulations. The "corporate price", that is the price established by the corporate organisations, was in fact "the right one" because it satisfied the interests of all: so, the setting of prices was delegated to commissions made up of representatives of the winegrowers, winemakers, wine merchants, as well as those arrived from the Unione Italiana Vini (literally Italian Wines Union) and other undefined "collective entities having relevant links to the wine sector" (Benatti and Castelletti 2022). In 1938, in a context of an asphyxiating bureaucratisation dominating Italian economy, all viticulturists (landowners and tenants and sharecroppers) were incorporated into the provincial producers' consortia, while the provincial viticulture consortia were transformed into provincial viticulture sections. However, the new structures were short-lived: in 1940, the provincial producers' consortia became economic agricultural bodies and the provincial wine-growing sections depended on them (Annuario 1941–42). Finally, since 1944, in regions where allied army arrived, these institutions were abolished and incorporated into the new provincial agricultural statistics offices.

The Fascist intense season of legislative measures on oenology was accompanied by a substantial stagnation of the wine sector. The Fascist regime's hammering pressure to allocate efforts and resources to increasing cereal production contributed significantly to the downsizing of the viticulture, which was also penalised by the fall in exports and prices, and the restriction caused by the autarkic policy to the foreign chemical products that were essential for fighting phylloxera and other plant diseases. Besides, the tightening of the contractual conditions of the agrarian contracts led peasants to neglect the production of good wine from the landowner's grapes. So, during the 1930s there was a marked impoverishment of the wine sector.

After 1945 public intervention attitudes changed and renounced to regulate every aspect of the wine sector (Annuario vinicolo d'Italia 1962–63). New Italian legislation in particular regarded the use of colouring agents, flavourings, the repression of fraud, impurities and manipulations creating a danger to people's health. Fraud and wine manipulation were in fact numerous. Although there existed measures in place to prohibit the adulteration of wines on the market, there were numerous itinerant vendors and small producers selling outside of any regulation. Consumption was low and the large quantities of wine sold off-premises to avoid paying duties and municipal taxes did not stimulate to improve the quality of ordinary Italian wines: so, most of them had a poor quality.

Only certain special wines, such as Moscato di Pantelleria and Marsala, were regulated. It was established that the designation Moscato di Pantelleria was reserved exclusively for wine obtained from the zibibbo grape (muscatellone), suitably dried and produced in the territory of the island of Pantelleria and then made in the same isle with the sole addition of ethyl alcohol. Concerning the Marsala, an exclusive production area was delimited between the Sicilian provinces of Trapani, Palermo, and Agrigento.

Furthermore, Italy signed important agreements with Germany, Switzerland, and France to encourage wine trade. In the 1949 agreement with France, for the first time a long and detailed list of typical Italian wines that could be exported to France was drawn up. So, in a national context where the denominations of origins did not exist, this last was used to defend the claims of Italian wines sold abroad. As it had happened at the end of the nineteenth century, it was the need to participate in the international trade that gave Italian wines a precise identity, improving their quality and recognisability.

While the first effects of Rome Treaties creating the European Common Market (1957) were arriving in the six member countries, Italy, France, Spain, Portugal, Cuba, Israel, Romania, Czechoslovakia, and Hungary also signed the Lisbon Agreement on Designations of Origin (1959). The purpose of the agreement was to protect designations of origin as effectively and uniformly as possible, meaning the geographical name of a country, region, or terroir designating a product which originated from it and whose qualities and characteristics were due exclusively or essentially to the geographical environment, including natural and human factors. The signatory countries in particular agreed on the creation of a vineyard register and the introduction of new measures to regulate the wine production and trade. The agricultural area devoted exclusively to vineyards started to increase. In 1960, the list of names of typical Italian wines was communicated to other countries belonging to the European Common Market: in the same year a proposal for a law to regulating designations of origin was presented in the Italian parliament.

Starting with the 1960s technological innovations improved the quality of the Italian wine, in particular those concerning new conservation methods and packaging techniques. During the 1960s, the reduction of the dimension of the barrels (the *barriques* in wood oak) giving more oxygen during the fermentation or the recourse to new giant tanks (in stainless steel or fibreglass) to better conserve the product allowed producers to create new different types of wine having a better average quality and also a more competitive price. Moreover, the introduction of new technology for bulk wine transport also affected the quality of wine sold on international markets. In particular, the introduction of new packaging technique

"Flexitank" (big bag-in-box with a capacity of 16,000–24,000 litres) progressively substituted steel containers. The quality attributes of wine were not deteriorated because of reduced oxidation: a better preservation of organoleptic characteristics was guaranteed. So, by Flexitank it was possible to transfer wine from the areas of production to all areas of consumption. Besides, it was possible to significantly reduce freight costs and to sell everywhere high-quality wine at a competitive price. This obviously increased competition on the international markets because more quality wines were available. Other positive effect for consumers was the reduction of environmental costs for the wine transport. However, the glass industry had negative effects and an augmentation of unemployment was experimented in that industry as well as among people operating in wineries which were not able to resist in a market where the average wine quality increased and the related sale prices diminished.

In addition to packaging advances, different innovations related to the wine traceability also characterised the evolution of the quality characteristics of wine products. Among these, the Quick Response (QR) code was recently adopted within the food industry as a two-dimensional barcode (Tarjan et al. 2015). The QR collected a higher quantity of information than the one-dimensional code, and it could be incorporated into users' smartphone applications: these latter allowed consumers to scan and decipher product information. The introduction of QR technology led to an increase of product knowledge and to a reduction of market failure associated to the information asymmetry between producers and consumers: this was in particular important for experiential products, like wine (Wilson and Quinton 2012).

Organisational innovations in the wine sector were mainly related to changes in the organisation of the wine supply chain. Among these, the major variations depended on: the implementation of sustainability aspects within the supply chain; the adoption of voluntary safety and quality standards; the role of producers and retailers in the management of vertical relationships. As sustainability of wine production progressively increased its relevance in the wine market and played a strategic role at international level different initiatives were developed to promote the sustainability of wine supply chain. The first was launched in California in 1992 (Integrated Pest Management Programme). Later, many countries started to support sustainability, especially countries belonging to the "New world of wine", such as "California Sustainable Winegrowing Program" (California, United States), "Entwine" (Australia), and "Sustainable Winegrowing"

(New Zealand). These initiatives brought to the implementation of voluntary standards for environmental, social, and economic sustainability of the wine production. Some initiatives related to sustainable-related wine supply chains also concerned the European terroirs. For example, the "Vignerons en Développement Durable" programme, a French collective brand for the sustainable viticulture based on the subscription of regulations composed by responsibilities with the aim of reaching goals connected to the sustainability.

Another example was "VIVA Sustainable Wine" which was developed by the Italian Ministry of the environment with the collaboration of many Italian Universities and research centres. Its aim was the evaluation of the company's performance from the environmental, social, and economic perspectives. It also considered the communication tools used to reach the final consumer, in particular a QR code identifying the company's results with reference to four indicators: Air, Water, Vineyard, and Territory.

It was also important the adoption of voluntary schemes which were referred both to public and private standards. Public standards related to the recognition of wine origin through the PDO and PGI certifications and environmental-friendly wine attributes through Organic standard. Private standards were linked to holistic approaches to renewable agriculture through bio-dynamic procedures (Demeter) or to the adoption of standards reducing unfair practices among winemakers. Such rules referred to traceability schemes entailing a higher complexity compared to the mandatory scheme introduced by Regulation 178/2002. In specific, these standards, like ISO 22005, referred to traceability standards, whose system had a high level of depth, breadth, and precision. Traceability depth referred to the sectors of the wine supply chain which were involved by the system. The breadth of the system referred to the amount of information traced. The precision of the traceability referred to the probability to reconstruct the complete history of a certain product and to the dimension of the tracking unit used to trace products. The higher the breadth, depth, and precision of traceability, the higher its complexity, and the higher the probability of efficiently managing unfair practices and exogenous shocks within the wine supply chain (Wu et al. 2012; Manning and Soon 2014).

Changes also concerned the organisation of supply chain driven by these voluntary systems: an increase in transaction transparency and in the bilateral dependency of economic agents was revealed (Banterle and Stranieri 2008). Indeed, the introduction of these voluntary standards increased the supply chain efficiency due to a strengthening of vertical relationships and the reduction of transaction information asymmetry. However, the adoption of complex traceability faced some difficulties, which were associated to the costs for its adoption, the type of product considered, and to the complexity of the supply chains. Moreover, the increase of transparency offered by complex traceability caused opposing effects within the supply chain, because of the presence of different interests among food and drinks firms and their tendency to behave opportunistically during transactions (Ringsberg 2014). The decision on the voluntary traceability for wine producers in fact depended on the company's strategy, that is from its strategic incentives towards the implementation of traceability.

Other relevant changes involved the role of producers in the organisational innovation of the supply chain: an increasing role of the winemakers' associations was progressively registered. Such forms of supply chain organisation implied an increase in supply chain coordination due to the integration of the production and processing phase. Moreover, a strengthening of vertical relationships was due to the introduction of supply chain agreements implying new severe production rules to be respected by all members of the associations involved. Retailers also played an important role in the reorganisation of wine supply chains. The strategic role of wine retailers within the supply chain depended on different aspects: their strategic position at the end of the supply chain and their big dimensions compared to wine producers.

With regard to the first aspect, the direct connection with consumers allowed them to quickly perceive their preferences and needs. This also allowed to reach information about market changes more quickly than the other actors of wine supply chain and to have more available information during negotiations. Moreover, the retailing was characterised by some big companies which concentrated a high percentage of food supply. On the opposite, most of the wine producers had small dimensions. For example, in Italy the system of wine production was based on about 55,000 operators subdivided into producers-winemakers, winemakers and winegrowers' associations.

The first and the last category of companies were examples of supply chain integration, whereas the second one was part of supply chains mainly organised through hybrid forms of transacting, such as contracts and similar agreements. More than 90 per cent of Italian wine companies was represented by winemakers even if they produced only a quarter of total national wine production. This entailed a power asymmetry between retailers and most of the agents of the wine supply chain and a progressive affirmation of retailers' power: these latter were the leader of the wine supply chain, coordinating the activities of the other agents.

Retailers centralised information and production flows of the supply chain in order to better monitor activities and to guarantee a higher degree of food safety and quality. To reach this goal, they introduced private standards with the aim to standardise quality procedures within the food supply chains. British Retailers Consortium (BRC) and International Food Standard (IFS) represented two relevant retailer standards for the efficient management of the supply chain. In specific, BRC was introduced by retailers in order to standardise the rules for suppliers with regard to food safety, food quality and other parameters (Contato 2007). This standard also introduced rules related to environmental and social sustainability. The environmental aspects related to a reduced use of chemicals in production processes, and to an efficient waste and water management within the food supply chain. The social aspect of this certification was based on the respect of work conditions with regard to labour rights and work safety issues.

The effects of technological changes were strengthened by new rules, in particular the European Economic Community Regulation No. 24 of 4 April 1962: it in fact established the birth of Italian "vini di qualità prodotti in regimi determinati" (VQPRD, that is quality wines produced under specified regimes) and then the new Law No. 930 of 12 July 1963 classified Italian wines as "Vino da tavola" (VDT, that is wines for table) and "Vini di qualità" (quality wines) which were DOC or DOCG (that is with a controlled denomination of origin or a controlled and guaranteed one). Before the promulgation of the law on designations, wine consortia and producers filed many trademarks: from 1950 to 1962, 32 trademarks were registered referring to Chianti in its different varieties (superiore, stravecchio, da tavola, secco). Many of the labels registered were written in English (Red Chianti Wine), specifying that the wine was for export: so, winemakers autonomously and spontaneously prepared their products before the introduction of the new law. However, the first wines having the denomination of origin arrived in 1966 only: there were discussions concerning the changes of the production procedures in a social context which remained linked to the oenological tradition and the related idea that only old productive systems guaranteed the quality. In a few years, the Italian wine sector underwent a strong transformation: it was normal to have resistances, moreover among the small wineries and the old winemakers. Besides, the legislator's intentions tended to enhance the value of the most valuable production, but many wines did not have sufficient quality standards to be classified as DOC. Only if they achieved the minimum characteristics, they laid down in the specifications. In the same classification, therefore, there were wines of great value alongside others of much lower quality. So, many producers of high-quality wines preferred to keep their wines in the generic table wine classification. Please note that the shortcomings and inefficiencies were overcome only with Law No. 164 of 10 February 1992 "New Rules for Designations of Origin": the classification of wines continued to follow the rules of the previous law, but there was a greater selection based on quality allowing to better identify the vines and the terroir.

Another decisive step allowing a relevant renewal of Italian winegrowing and winemaking was represented by the provincial plans for the development of viticulture, drawn up by the agricultural inspectorates in accordance with the law of 2 June 1961, that is the first "green plan" or five-year plan for the development of agriculture. The plan provided for the granting of subsidies to agricultural operators, preferably to individual or associated farmers, for the improvement and development of high value added rural sectors such as olive-growing, citrus growing, fruit growing in orchard, and viticulture. The renewal programme continued with the approval of the second "green plan" (Law No 910 of 27 October 1966), which granted subsidies for the planting of specialised vineyards and vine nurseries. So, using public investments, an ambitious programme was drawn up to help landowners who decided to only operate in the viticulture: ally, subsidies were established. The plants eligible for the contributions had to belong to the following categories: cooperative specialised vineyards, worked by viticulturists who had joined their forces to introduce improvements in wine-growing and winemaking; specialised vineyards belonging to farms, suitable for the use of mechanical means and consequent reduction of workforce; vine nurseries, cultivated by associated farmers or by reclamation consortia operating in terroirs where valuable vines were grown; all other local vines considered of particular interest.

Planned works had to comply with various agro-technical requirements. Improvements had to be carried out on land previously broken up and landscaped, including the water regimentation. Another evaluation criterion was represented by the adoption of rootstocks having an undoubted affinity with the chosen vine and were suitable for the different types of soil. Specific attention was paid to the choice of vines, preferring for the white and red grape varieties best suited to the characteristics of the terroir. In addition, the distance between the rows of vines had to allow easy mechanised cultivation. Furthermore, for the training of the vines, the horizontal cordon system (such as Cazenave and Sylvoz) or the vertical system with multiple stakes had to be preferred, without excluding other methods guaranteeing the greatest possible productivity. The green plans significantly contributed to changing the Italian viticulture: the land surface occupied by specialised vineyards increased and there was a marked increase in the production of grapes in specialised crops.

At the same time, many designations of origin were established, sometimes for political reasons: they were often referred to wines that did not have a high quality and did not enjoy any kind of notoriety or prestige in the wine market. Moreover, these wines did not have the requisites justifying a real high price: thus, many of these designations brought the institution of designations of origin into disrepute. Furthermore, many of them had a small production base and producers did not have the resources to carry out any kind of promotional activity to build up their image and facilitate their sales on the wine market which progressively became more competitive. In 1992, of the 226 existing designations of origin, 143 produced less than 15,000 hl. The solution to overcome this excessive fragmentation was the creation of regional DOC, but it was not very successful because wine producers were against it: in Tuscany, for example, they argued that a regional name could be used to designate IGT wines that is wines having "Indicazione Geografica Tipica" (literally "Typical Geographical Indication"), but it was not possible to create a "Tuscan DOC" wine including the entire regional territory because too many different terroirs existed ("Il corriere vinicolo", n. 25, 24 June 1996, p. 2).

This wide productive diversification explains why in the new millennium, with its 863 denominations—299 concerning food (35 per cent), 526 regarding wine (61 per cent), and 38 relating to spirits (4 per cent)—Italy overcomes France (764) and Spain (358). Moreover, the most important period for the creation of geographical indications of origin was exactly after the discussion about the exaggerated use of the denominations of origin, that is the years from 1996 to 2012. While those concerning predominated in Emilia-Romagna (45), Veneto (38), and Lombardy (36), those regarding prevailed in Piedmont (59) and Tuscany (58). A striking result for a country which, in the 1960s, had none.

The surface area of vineyards indicated in the DOCG register was 45,000 hectares in 2005 (Unioncamere 2007): in the early new millennium, it yearly increased by 5.6 per cent. In Tuscany (almost 25,000 hectares) and Piedmont (almost 15,000 hectares) existed about 88 per cent of the land dedicated to DOCG wine production. Campania and Umbria were the regions registering the average annual increase (respectively 38.3 and 29.1 per cent). Abruzzo also had a strong growth, but the recognition of its DOCG arrived later: for example, the Montepulciano d'Abruzzo DOCG was born in 2003 only.

In 2005 more than 3.7 million quintals of grapes were harvested for DOCG wines, with an average annual increase of 6.7 per cent compared to the harvest in 2000: the most significant harvests were obviously registered in Tuscany and Piedmont. In 2005, there was a production of about 11 million hectolitres of DOC wine, with an average annual increase of 2.9 per cent. The regions contributing most to DOC wine production were Veneto (21 per cent of national production), Emilia Romagna, and Piedmont. There were also a marked high annual increase in production: Liguria (+17.6 per cent), Basilicata (+15.5 per cent), and Tuscany (+13.1 per cent). In opposite, there was a significant drop in production in Friuli (-4.9 per cent per year).

In the same year, there was a production of almost 2.2 million hectolitres of DOCG wine, with an average annual increase of 4.2 per cent compared to 2000. Tuscany produced about 1.2 million hectolitres (54 per cent), and Piedmont followed with more than 370,000 hectolitres (17 per cent). Strong increases in production were recorded in Campania, Veneto, and Umbria, albeit with considerably lower production, in quantitative terms, than the two leading regions Tuscany and Piedmont (Unioncamere 2007).

2.5 The Influence of European Regulations

It is not possible to study Italian viticulture since the 1960s without referring to EU agricultural policies. The quality in the European wine production and market was in fact influenced by new technologies but also by new European rules. These latter progressively modified the rules concerning the distribution of all European products, the wine included: please note that the adjective European concerns the EEC country members (Italy was one of the founders) and so its extension varies following the EEC enlargements. The Common Agricultural Policy (CAP) was born in 1962 and in 1964 it started to regulate the methods of winemaking and the characteristics concerning all the European wines (Tedeschi 2011). The European market for the wine sector was organised in order to enable a gradual convergence of prices and to eliminate customs barriers: it established a single internal market with one common tariff for the rest of the world. So, European institutions progressively decided the evolution of the European wine production and market and in particular the relevance of the wine quality. They in fact established all rules concerning: the elimination of tariffs protecting national products and the actual level of competition in the European wine market; the chemical products which it was possible to use during the winegrowing against fungal diseases and insects; the minimum alcohol content for wines of member countries (it was higher for Italy and France, lower for West Germany and Benelux where 5 degrees were enough); the percentage of preservatives as sulphites; the contents of labels which wine producers had to put on their bottles, and more.

These European rules also concerned the recourse of new technologies allowing a relevant improvement of the yields in grapes and, moreover, of the quality of the wine (e.g. thanks to new fermentation process and modern types of casks favouring a better conservation of the wine). Besides, the evolution of the rules about the origins of the products and the related labelling strongly influenced the wine producers' decisions concerning the characteristics of the wine they wanted to make and sell. Face to the new world globalisation of the wine market, a new European wine policy was shaped with the aim to improve the average quality and to promote a quality differentiation, that is to create more types of wine and encounter all target consumers. The CAP progressively enlarged to all European countries the best rules existing in France and Italy, that is the best world wine producers and, moreover, the countries where the wine sector has a great relevance for both the national agri-food system and the preservation of the territory (Meloni and Swinnen 2013, 2018; Meloni et al. 2019). So, new labels such as CDO (Controlled Designation of the Origin, sometimes Guaranteed too), PDO (Protected Designation of Origin), PGI (Protected Geographical Indication), and, more recently, "organic" were written on the bottles of European wines to indicate to consumers how and where the wines they drank were made.

The respect of new labels, which progressively became more severe, obliged European winemakers to improve the quality of their products to maintain their niches of market: the final result was (and it continues to be) positive and the flavours of new quality wines were able to encounter the consumers' tastes and enlarge the wine market. New European rules also concerned organisational innovations, in particular the wine supply chain. Relevant changes in fact regarded the implementation of sustainability aspects within the supply chain and the role of wine producers and retailers: the aim was to improve the quality of the European wines and the respect of the environment.

While in the previous decades, the CAP and national public authorities justified their subsidies to the viticulturists with the need to avoid that young people left the European countryside, since the 1990s and in particular in the new millennium the aid to the winegrowers started to depend on the positive effects, for the environment and also for the tourism, of the planting and cultivation of vines in the European countryside. European rules influenced the wine producers, retailers and consumers' decisions: however, also the evolution of the wine market obliged European institutions to adapt rules to changes provoked by those decisions.

European institutions had to balance the EU citizens' exigencies for more wine with a good quality (higher) and price (lower) ratio with the necessity of the wine sector to remain profitable and maintain job places as well as the conservation of the environment and the increase of the landscape value granted by the viticulture. Besides, European institutions had to face the negative trend in the wine demand: this last depended on the main wine markets and was mainly related to the changes in consumers' lifestyles and related diets. In the 1960s and 1970s, the consumption of wine in the main European winemaking countries was part of the everyday diet because it was an important part of the caloric intake necessary to perform work activities. From the 1980s, the wine became a matter of occasional consumption. The growing attention to health-related aspects, the increasing number of sedentary lifestyles and the growth of alternative (alcoholic or not) drinks in fact led to a decrease in the demand of wine (Hertzberg and Malorgio 2008). Moreover, the negative trend concerned the low-quality wines: consumers progressively reduced their consumption, but they drank better wine. This increased the competition in the wine market and stimulated winemakers to further improve the quality of their products and to better analyse the consumers' preferences towards wine quality attributes. Face to these changes, the relevance of labelling grew up: so, European institutions had to modify some rules and in particular increased sanctions against wine producers' false statements.

Due to the great importance of wine sector for the whole European agricultural economy, European regulation also provided a wide range of rules to protect and differentiate the wine production: this influenced European countries producing wine but also countries where the wine market progressively assumed a social and economic relevance (Gaeta and Corsinovi 2014; Harvey and Waye 2014). The main changes characterising the wine sector were related to the product quality differentiation and they concerned both processes, logistic along the supply chain and the market recognition of quality attributes of wines through labelling (Banterle and Stranieri 2013).

The first European relevant rule arrived at the beginning of the 1970s: it was Regulation (EC) No. 817/70 that introduced a specific regulation for the provisions relating to "Quality wines produced in specified regions" (QWPSR). A quality wine produced in specified regions could be sold under the name of the region: it was the producer member country to guarantee its origin. Main examples of this regional "recognized quality" were: "Naturwein", "Originalabfullung", "Spatlese", "Auslese", "Beerenauslese", and "Trockenbeerenauslese" for German wine; "Champagne" for French wine. Besides, these names could be followed by "recognized expressions of quality" as: "Qualitatswein" in Germany, "Appellation d'Origine Contrôlée" and "Vin délimité de qualité supérieure" in France, "Denominazione di origine controllata" and "Denominazione di origine controllata e garantita" in Italy, and since 1986 "Denominación de origen protegida" in Spain. European institutions used and improved the existing rules: they copied the French "Appellation d'Origine Contrôlée" (AOC, literally controlled denomination of origin, CDO) which was the first quality label as it was introduced in France for the wine sector in 1935. They diffused a new quality scheme based on the name of a country, province, or terroir and designed a wine whose characteristics were exclusively or mainly due to its geographical origin.

Among the distinctive features of these products were also included the characteristics of human capital and natural resources: this progressively became for consumers a guarantee of the quality of the wine. However, wines without the indication of origin were not necessarily of low quality: some winemakers could in fact realise excellent products using a "bouquet" of grapes arriving from different European terroirs. Furthermore, many winemakers considered that the name of the winery was the most relevant "indicator" of the real quality of the products: some others, who were evidently very critical of the procedures for checking compliance with the local rules in order to obtain the CDO, pointed out that the latter actually helped the worst producers—the best ones did not need it.

In addition to quality recognition policy, regulations also introduced boundaries on the replanting of vines for European countries. Regulation 1161/76 introduced new rules on the definition of intrinsic quality parameters for wine. In specific, such normative framework aimed at introducing and changing rules on different aspects.

Firstly, each European country had to fix a minimum natural alcoholic strength for each of the quality wines produced within its territory. Secondly, winemaking and processing methods adopted for quality wines had to be defined. Third, it was suggested that permission for the sweetening of a quality wine had to be asked to a Member State. The same authorisation was also asked for the enrichment, acidification, and de-acidification methods. With regard to planting restrictions, a complete ban on all new plantings for table wines was introduced: European institutions aimed at limiting the production of "generic" wine and, at the same time, incentivising the production of quality differentiated products.

Regulation (CEE) 823/1987 introduced the first system of European wine quality recognition. Such normative framework aimed at homogenising the wine quality policy of each member country and it regulated conditions of production and characteristics for all QWPSR. With this Regulation new quality schemes were introduced:

- Quality wines produced in specified regions
- Liqueur wine quality produced in specified region
- Sparkling wines produced in specified regions
- Semi-sparkling wine quality produced in specified regions

So, winemakers adopted a disciplinary of production, where the following information was provided: the determination of the production area; types of grape; cultivation methods; winemaking methods; title minimum blood alcohol; yield per hectare; analysis and assessment of organoleptic characteristics. Italy in particular implemented such rules with the national law 164/92. Different quality labels were introduced within the Italian territory to diversify the quality characteristics of wines. They were DOCG (Denominazione di Origine Controllata e Garantita), DOC (Denominazione di Origine Controllata), and IGT (Indicazione Geografica Tipica). For the consumers, they became the main quality labels introduced for wine quality differentiation.

The label DOC was attributed to wines produced in limited geographical areas (usually small/medium sized) and made following strict rules, which included: the use of some defined types of grapes and some established winemaking techniques; the existence of predetermined wine characteristics; the sale only after accurate chemical and sensory analysis. For these wines was also permitted some particular designation: "Classico" (classic) if the wine was produced in terroirs where the viticulture has existed since ancient times; "Riserva" (reserve) if the wine underwent a period of ageing (two years or more). At the same time, some restrictive product specification obliged the producers to the following requirements: the DOC designation on the label; the boundaries of the production area territory; the maximum yield of grapes per hectare; a minimum alcohol volume; the specification of chemical, physical and organoleptic characteristics of the wine; the production conditions (climate, soil, altitude, soil exposure); the authorised vine types; the density of the installations, pruning systems, and so on; the chemical and organoleptic examination mode; any minimum period of ageing in wood and bottle; any indication of the areas authorised bottling.

The DOCG was a particularly prestigious certification reserved for certain DOC wines of high quality or with a high international recognition. Producers had to follow rules more severe than those concerning DOC wines. These wines had to be marketed in containers/bottles of less than five litres and carry a label detailing the State, the guarantee of origin, the quality, and the number of bottles produced. To obtain the market benefits granted by this particular designation, wine producers' costs increased: so, only the best and expensive wines received it and this justified the best customers' preferences, that is the diffusion in the high-quality wine market (Belletti and Marescotti 2007).

The IGT was quality awarded to table wines, which generally had a quite large production area and usually had a quality inferior to DOC and DOCG wines. IGT wines corresponded to the French "Vin de Pays" and the German "Landwein". The IGT wines production was regulated by simple and flexible rules. The following information was required: the indication on the label of the origin and the names of grape varieties; the boundaries of the production area territory; the list of grape varieties used in the production; the colour and wine type; the maximum yield of grapes per hectare; the alcoholic volume; the grapevine yield; the authorised corrective practices.

Another important intervention for the regulation of European wine market was the Reform of 1999: it had the aim to align European supply and demand through the restructuration of large land surface dedicated to vineyards, to eliminate the use of intervention price for surplus productions, to arrange regional diversity, to recognise the role of best wine producers giving them the possibility to guarantee a production corresponding to demand existing on the market for high-quality products. To achieve these objectives, new rules in particular concerned production potential, market mechanisms, oenological practices and processes, designations of origin, presentation and protection of national products, and trade relations with third countries (that is no EU members). In addition to these provisions, new rules established a classification of authorised wine grape varieties, a wine inventory, and a vineyard register.

However, this reform was insufficient in reducing wine surpluses and European institutions had to pay too much money for wine sector. So, in 2008 the European institutions reorganised the European wine market starting from the 2003 CAP reform introduced by Regulation (EC) 1782/2003. The reform also followed Regulations (EC) 509 and 510 of 2006 establishing new severe procedures to obtain new European labels. It had different aims and the normative framework in particular recognised the importance of: the abolition of the ineffective public intervention in European wine market; the convergence between European wine production and demand; the increase of European wine producers competitiveness; the reinforcement of European wines reputation; the improvement of market share both in the internal market and outside EU; the protection of the traditions for European viticulture and the enhancement of the social and environmental role of winegrowing in rural areas.

With Regulation (EC) No. 479/2008 the oenological practices and the policy for wine quality were changed in order to harmonise the EU quality policy for food products with that of wine products. More precisely, such Regulation linked the PSR labelling normative with PDO and PGI rules (that is, Italian DOP and IGP). The new normative framework distinguished between wines of quality produced in a specific area and wines without a geographical indication. Within the first group, there were PDO and PGI wines. PDO referred to wine which was entirely produced and transformed in a given geographical area. PGI referred to wine products where at least one production step within the supply chain was based in a specific geographical area. In specific, for PDO, the requirement was that "the production must take place in the geographical area [and] cover all the operations involved, from the harvesting of the grapes to the completion of the wine-making processes, with the exception of any post-production processes". For PGI wines, "the maximum 15 per cent share of grapes could arrive from outside the demarcated area, but it had to originate from the Member State or third country in which the demarcated area was situated". The consequence of this new regulation was that many IGT wines (or equivalent in other EU countries) became PGI ones. The result was a quality upgrade for many wine products.

However, the homogenisation of the quality requirement for wine products with other foodstuffs led to a risk of consumer confusion towards new labels. Besides, it was difficult to actually protect European wines in international markets and this aspect was at the core of political debate in main European wine producers' countries and in particular in Italy (Chiodo 2008). PDO and PGI wines in fact represented (and they continue to represent now) a strategic element of Italian agri-food system: a great part of the Italian countryside (arriving at almost 50 per cent) was in 2013 dedicated to the viticulture and from 50 to 80 per cent of the wine production had the label PDO clearly demonstrating the specialisation of Italian wine production towards high quality. Even if some best winemakers continued to realise excellent wines without following the rules for the PDO wines (the excellent flavour of their wines in fact depended on the mixing of grapes and must arriving from different terroirs) the relevance of European labels for consumers progressively led most of European wine producers to improve in quantity and quality the vines dedicated to the production of PDO wines. So, the improvement of the quality concerned most European wines and this increased the competition in the wine market and, consequently, obliged European institutions to intensify the controls concerning the respect of rules.

In 2012, an important and definitive step concerning the European "organic" wine was also reached. Even if in 1991 the first European rules concerning the production of grapes using "organic agriculture" (CE n. 2092/91) were promulgated, no new rules had in the following years regulated the "organic" winemaking. It was indicated which and how many plant protection products as well as fertilisers could be used: the necessary control mechanisms were established. However, the production of "organic" wine was not regulated and it did not exist a related European label. This happened because the European rules actually considered all organic products and not only the wine sector. The effects obviously concerned Italian wines too: it was impossible to use the label "biologico" (literally "organic") to indicate wines made from grapes grown in accordance with principles of organic farming which typically excludes the use of artificial chemical fertilisers, pesticides, fungicides, and herbicides.

New regulation (EU 834/2007) indicated that since 1 January 2009 it was possible to produce "organic", but it did not explain the procedures for the related winemaking: strong discussions in fact divided wine producers and EU member countries, in particular about the use of sulphites. In June 2010, no agreements were signed and paradoxically the first real rules were drafted by private organic winemakers associations coming from France, Italy, and Spain (and also Switzerland) and creating the European Organic Winemaking Carta (EOWC). This last in fact became the basic document for the European rules (EU 203/2012) establishing the actual birth (on 1st August 2012) of the first European wines having the label "organic". After new long discussions, it was finally found a compromise: new label depended on the residual sugar and the quantity of sulphites remained linked to what happened to their wines. Italian and Spanish winemakers considered this solution as very negative (or better "not quite ambituos enough"): however, the position of Germany and Austria protecting their wine producers did not allow a real alternative and previous experiences showed that it existed the risk of prolonging the lack of rules for many more years.

In 2013, the European institutions adopted a further reform with the aim at harmonising and simplifying the CAP outlines: the regulatory frame of the 2008 reform was preserved but, for the wine sector, some changes were introduced and in particular concerned designations of origin and geographical indications. As labelling progressively assumed great relevance in the wine market, the labels became more severe and, at the same time, were more related to the respect of the guidelines established by consortia producing PDO and IGP wines. Besides, new rules definitively illustrated how to produce an "organic" wine improving the procedures in favour of the respect of the environment and reducing sulphites.

Moreover, new rules were part of the wider EU intervention concerning the CAP for the years 2014–2020 (Ashenfelter et al. 2018). So, other relevant topics which were regulated concerned the national support programmes and the scheme of authorisations for vines plantings. European institutions considered the wine sector relevant for the environment as well as for the economy: a correct evaluation considering that two years after, in 2015, the European wine sector granted 3 million direct full-time jobs and the market value of European wine overcame 100 billion of euros (Eurostat 2015). So, these actions were regulated: the restructuring and conversion of vineyards; the green harvesting; the mutual funds; the harvest insurance; the allowed instruments and productive systems and the related investments; by-product distillation; the promotion in third countries. By this normative framework European institutions wanted to foster new products and processes development, especially related to the introduction of advanced systems of sustainable wine production. In addition, it promoted the spread of information communicating with consumers about the responsible consumption of wine and about the system of the designations of origin and geographical indications.

With regard to authorisations, the planting rights approach was abolished by December 2015. New personal authorisations were granted without charge and were not transferable. So, in 2016 a new system for the management of vine plantings was set up as the "Scheme of authorisations for vine plantings" in which EU Member States made available each year authorisations for new plantings corresponding to "1 per cent of the total area actually planted with vines in their territory, as measured on 31 July of the previous year". This plan was based on the outcome of the High Level Group on Vine Planting Rights organised in 2012: its full realisation was foreseen for 2030. Other regulatory interventions succeeded in order to strengthen the quality of wine. They referred to three main issues: the market recognition of wine quality through the introduction of new labels; the introduction of rules for planting restrictions; and the regulation of intrinsic quality wine attributes and their production methods.

All new rules promulgated in the new millennium gave great attention to the respect for customers' rights. European institutions increased controls on wine producers' statements influencing wine consumers' choices: consumer involvement and product knowledge in fact affected consumer's preferences (Barber et al. 2009). The incorrect information on the wine quality and origins was in fact able to modify the market in favour of not honest wine producers. European institutions also knew that some extrinsic attributes could be modified without changing the product itself (e.g. the price, packaging, labelling, and brand name) as well as intrinsic attributes were directly connected to the product, to the processing method, and to the consumers' perception of it (e.g. the alcohol content and product sensory characteristics). The first ones concerned marketing only and

European institution had to avoid false information misleading consumers to the detriment of honest wine producers' competitiveness: the second ones were strictly related to the protection of consumers' health and of honest wine producers whose sales could be reduced by the fear related to the presence of adulterated and dangerous wines on the market.

New laws were de facto based on new studies concerning objective traits, sensorial variables, and reputational attributes of wine producers, all aspects affecting the consumers' preferences: so, they were interesting for European institutions protecting European customers. Objective traits were (and obviously continue to be) the price, origin, denomination of origin, grape variety, name of producer: the price of wine in particular influences buyers' choice when the other characteristics of the product are not known by consumers, or when it is difficult for these latter to evaluate the quality of the product. The other objective traits become relevant when consumers have a major knowledge of the characteristics of the terroirs and the wines local wineries are able to produce: it is evident that a false information concerning these aspects deceives customers and modifies the correct competition in the wine market.

On the contrary, the sensorial variables include product characteristics which are not known by consumers before the purchase: it is in fact possible to discover them only after the product tasting (e.g. wine's arome, body, finish, and harmony of components); when these aspects are different from consumers' expectations, it is evident that sales decrease in a short time, so the European institutions' interest concerns only cases in which differences are provoked by conservation defects (and the possible negative effects on consumers' health) or by false claims on labels (wine producers indicate different characteristics stripping customers).

Finally, reputational attributes are represented by all expectations about wine quality, built up through past experiences with the producer, the brand and the designation of origin: they are linked evidently to sensorial variables and European rules have to avoid that false statements deceive consumers and also create problem markets for honest winemakers operating in the same terroir (Landon and Smith 1997; Harvey et al. 2014).

In the new millennium, European rules concerning the wine market also considered the problem of the sustainable development: consumers in fact progressively showed an increasing interest in sustainable products and so sustainability has become one of the primary concerns for winemakers. Sustainability attributes in fact attracted consumer's attention as the other attributes: so, wine producers progressively modified their strategies concerning the quality differentiation of their products trying to underline the respect for the environment during the winegrowing and winemaking. This evidently implied that European rules had to regulate and control wine producers' statements and, moreover, verify that information was clear and did not confuse consumers in their purchasing decisions: the risk of information overload in fact exists and the choice of labelling only the most important information helps wine producers to effectively differentiate their products from competitors (Stranieri and Tedeschi 2019).

All these aspects explain why the European wine sector was deeply regulated since the 1960s: European regulations de facto influenced international markets too as almost half of the world's vineyards were located (and continue to be) in the European countryside (e.g. the vineyards surface in 2015 was 3,362,000 hectares, that is more than the dimension of Belgium). Moreover, the European wine producers continued to represent the main importers and exporters in worldwide wine markets: for example, in 2015, they produced the equivalent of 23 billion bottles, that is 63 per cent of the world wine production (Eurostat 2015).

European institution also had to consider the increasing degree of concentration of wine demand worldwide increased. In the first decade of the new millennium the highest levels of individual consumption remained concentrated in European countries with a strong wine tradition, such as France, Italy, and Spain: they in fact produced 80 per cent of European wine and 50 per cent of the world one. Other European wine producer countries, for example, Germany and Greece, continued to produce and consume limited quantities of wine, while Northern European countries without tradition in wine production revealed quite positive trends concerning the wine consumption. Outside Europe, the consumption also grew up and some countries registered a considerable increase in the demand for wine: this in particular happened in the United States, Russia, Australia, South Africa, Argentina, Canada, Brazil, Chile, and Japan. The related economic relevance for the European wine export implied that European institutions had to control, promote and verify the quality of European wine: at the same time they were committed to keeping tariffs low in the international markets, that is in a context where the wine sector represented only one of the European productive sectors and not one of those that guaranteed more income and jobs. This explains the Italian winemakers' feeling of not being sufficiently protected and of having to work alone on continuous quality improvement to always remain competitive.

References

Ainsworth, G.C. 1981. *Introduction to the History of Plant Pathology*. Cambridge: Cambridge University Press.

Alston, Julian, Jim Lapsley, Lena Sambucci, and Dan Summer. 2018. United States. In *Wine Globalization. A New Comparative History*, ed. Kym Anderson and Vicente Pinilla, 410–440. Cambridge: Cambridge University Press.

Ambrosoli, M. 1997. *The Wild and the Sown. Botany and Agriculture in Western Europe, 1350–1850*. Cambridge: Cambridge University Press.

Annuario vinicolo d'Italia. 1962–63. Milano: Unione italiana vini.

Annuario vinicolo d'Italia, 1941–42. Milano: S.A. Pubblicazioni di Propaganda Vinicola "Enotria".

Aparicio, Gema, Vicente Pinilla, and Raul Serrano. 2009. Europe and the International Agricultural and Food Trade, 1870–2000. In *Agriculture and Economic Development in Europe Since 1870*, ed. Pedro Lains and Vicente Pinilla, 52–75. London: Routledge.

Ashenfelter, Orley, Olivier Gergaud, Karl Storchmann, and William Ziemba, eds. 2018. *Handbook of the Economics of Wine. Volume 2: Reputation, Regulation and Market Organization*. Singapore: World Scientific Books Inserire.

Atti convegno. 1884. *Atti del convegno dei produttori di vino tenuto a Rome nei giorni 18, 19, 20 e 21 febbraio 1884*. Rome: tipografia eredi Botta.

Augello, Massimo, and Marco E.L. Guidi. 2012. Educating the Nation: Textbooks and Manuals of Political Economy in Italy, 1915–1922. In *The Economic Reader. Textbooks, Manuals and the Dissemination of the Economic Sciences during the Nineteenth and Early Twentieth Centuries*, ed. Massimo M. Augello and Marco E.L. Guidi, 124–157. New York: Routledge.

Banterle, Alessandro, and Stefanella Stranieri. 2008. The Consequences of Voluntary Traceability System for Supply Chain Relationships. An Application of Transaction Cost Economics. *Food Policy* 33: 560–569.

———. 2013. Sustainability Standards and the Reorganization of Private Label Supply Chains: A Transaction Cost Perspective. *Sustainability* 5: 5272–5288.

Barber, Nelson, Christopher Taylor, and Sandy Strick. 2009. Wine Consumers' Environmental Knowledge and Attitudes: Influence on Willingness to Purchase. *International Journal of Wine Research* 1: 59–72.

Barbero, Maria I. 2008. La historia de empresas en la Argentina. Trayectoria reciente y perspectivas. In *La nueva historia de empresas en España y América Latina*, ed. María Inés Barbero and Raúl Jacob. Buenos Aires: Temas.

Barrio, Patricia. 2008. El empresariado vitivinícola de la provincia de Mendoza (Argentina) a principios del siglo XX. *Historia Agraria* 45: 81–111.

Beardmore, Carol. 2015. Landowner, Tenant and Agent on the Marquis of Anglesey's Dorset and Somerset Estate, 1814–44. *Rural History* 26: 181–199.

———. 2019. *Financing the Landed Estate. Power, Politics and People on the Marquis of Anglesey's Estate, 1812–1854.* Cham: Palgrave Macmillan.

Belletti, Giovanni, and Andrea Marescotti. 2007. Costi e benefici delle Denominazioni Geografiche (DOP e IGP). *AgriRegioniEuropa*. https://agriregionieuropa.univpm.it/it/content/article/31/8/costi-e-benefici-delle-denominazioni-geografiche-dop-e-igp.

Benatti, Francesco Emanuale, and Paolo Castelletti, eds. 2022. *Dai vini tipici al testo unico. Profilo storico della normativa delle denominazioni di origine del vino in Italia (1904–2021)*. Milano: Unione Italiana Vini.

Betri, Luisa. 2002. Gli agronomi nell'Ottocento: dall'arte alla professione. In *Storia delle professioni in Italia tra Ottocento e Novecento*, ed. Angelo Varni, 173–184. Bologna: Il Mulino.

Bidolli, Anna P. 2001. L'istruzione agraria nella documentazione dell'Archivio centrale dello stato. In *L'istruzione agraria (1861–1928)*, ed. Anna P. Bidolli and Simonetta Soldani. Rome: Ministero per i beni e le attività culturali.

Bidolli, Anna P., and Simonetta Soldani, eds. 2001. *L'istruzione agraria (1861–1928)*. Rome: Ministero per i beni e le attività culturali.

Bigatti, Giorgio. 2000. *La città operosa. Milan nell'Ottocento*. Milano: FrancoAngeli.

Borras, Saturnino M. 2009. Agrarian Change and Peasant Studies: Changes, Continuities and Challenges—An Introduction. *The Journal of Peasant Studies* 36: 5–31.

Bowen, James P. 2010. A Landscape of Improvement: The Impact of James Loch, Chief Agent to the Marquis of Stafford on the Lilleshall Estate, Shropshire. *Midland History* 35: 191–214.

Brassley, Paul. 2000. Agricultural Science, Research and Education. In *The Agrarian History of England and Wales, vol. VII, 1850–1914*, ed. Edward John T. Collins, 594–649. Cambridge: Cambridge University Press.

———. 2005. Agricultural Education, Training and Advice in the UK, 1850–2000. In *Land, Shops and Kitchens. Technology and the Food Chain in Twentieth-Century Europe*, ed. Carmen Sarasúa, Peter Scholliers, and Leen Van Molle, 259–278. Turnhout: Brepols.

Carlucci, Michele. 1890. Concorso a premi per vini da pasto stabilito con R. decreto 10 ottobre 1885. In *Annali di agricoltura 1889*. Rome: tip. Eredi Botta.

Caron, Jean-Claude, and Frédéric Chauvaud, eds. 2005. *Les campagnes dans les sociétés européennes. France, Allemagne, Espagne, Italie (1830–1930)*. Rennes: Presses universitaires de Rennes.

Cartaña i Pinén. 2000, Jordi. 2000. Las estaciones agronómicas y las granjas experimentales como factor de innovación en la agricultura española contemporánea. Scripta Nova. *Revista electrónica de Geografía y Ciencias Sociales 69*(19). http://www.ub.edu/geocrit/sn-69-16.htm.

Cazzola, Franco. 1996. *Storia delle campagne padane dall'Ottocento a oggi*. Milan: Mondadori.

Ceccarelli, Giovanni, Alberto Grandi, and Stefano Magagnoli, eds. 2013. *Typicality in History: Tradition, Innovation, and Terroir / La typicité dans l'histoire. Tradition, innovation et terroir*. Bruxelles: Peter Lang.

Cecil, Robert. 1979. *The Development of Agriculture in Germany and the U.K. 1. German Agriculture 1870–1970*. Ashford: University of Reading—Wye College.

Chambers, Jonathan David, and Gordon E. Mingay. 1969. *The Agricultural Revolution, 1750–1880*. London: Batsford.

Cheneaux-Berthelo, Christiane. 2020. Science, art ou pratique? La place de l'agriculture dans les formes de transmission du savoir au XIXe siècle en France. In *Transmission et circulation des savoirs scientifiques et techniques*, ed. Barjot Dominique. Paris: Éditions du Comité des travaux historiques et scientifiques. Disponible sur: https://books.openedition.org/cths/13648.

Chiodo, Emilio. 2008. Vino: cosa cambia con DOP e IGP al posto di DOC e IGT. *AgriRegioniEuropa*. https://agriregionieuropa.univpm.it/it/content/article/31/15/vino-cosa-cambia-con-dop-e-igp-al-posto-di-doc-e-igt.

Cinotto, Simone. 2013. *The Italian American Table. Food, Family, and Community in New York City*. Urbana, Chicago and Springfield: University of Illinois Press.

Circolo enofilo italiano. 1892–1894. *Annuario generale per la viticoltura e la enologia, I–III*. Rome: tip. Bertero.

Ciuffoletti, Zeffiro. 2009. *Alla ricerca del 'vino perfetto'. Il Chianti del barone di Brolio. Ricasoli e il Risorgimento vitivinicolo italiano*. Firenze: Leo S. Olschki.

Conca Messina, Silvia A., Stéphane Le Bras, Paolo Tedeschi, and Manuel Vaquero Piñeiro, eds. 2019. *A History of Wine in Europe (19th–20th Centuries). Vol. I. Winegrowing and Regional Features*. London: Palgrave Macmillan.

Contato, R. 2007. The IFS (International Food Safety) and BRC (British Retail Consortium) Standard in Large Distribution. *Industrie Alimentari* 46: 180–183.

Cornu, Pierre, and Jean-Luc Mayaud, eds. 2007. *Au nom de la terre. Agrarisme et agrariens en France et en Europe du 19e siècle à nos jours*. Paris: Boutique de l'histoire.

D'Agata, Ian. 2019. *Italy's Native Wine Grape Terroirs*. Oakland, CA: University of California Press.

Da Silva Lopes, Teresa, Andrea Lluch, and Gaspar Martins Pereira. 2020. The Changing and Flexible Nature of Imitation and Adulteration: The Case of the Global Wine Industry, 1850–1914. *Business History Review* 94: 347–371.

Denis, Gilles. 2013. L'agronomie au sens large. Une histoire de son champ, de ses définitions et des mots pour l'identifier. In *Histoire et agronomie. Entre ruptures et durée*, ed. Robin Paul, Jean-Paul Aeschlimann, and Feller Christian. Marseille: IRD Éditions.

Depecker, Thomas, and Nathalie Joly. 2015. Agronomists and Accounting. The Beginnings of Capitalist Rationalisation on the Farm (1800–1850). *Historia agraria* 65: 75–94.

Depositi governativi di macchine agrarie. 1883. Rome: Tip. Eredi Botta.

Dewey, Peter. 2013. Landowner as Scientific Farmer: James Melson and Eynsham Hall Estate, 1866–1903. In *The Farmer in England 1650–1980*, ed. Richard W. Hoyle. London: Routledge.

Egerton, Frank N. 2012. History of Ecological Sciences, Phytopathology during the 1800s. *Bulletin Ecological Society of America* 93: 303–339.

Eurostat. 2015. *Vineyards in the EU—Statistics*. https://ec.europa.eu/eurostat/statistics-explained/index.php/Vineyards_in_the_EU_-_statistics.

Fabiani, Guido. 2015. *Agricoltura-mondo. La storia contemporanea e gli scenari futuri*. Rome: Donzelli.

Failla, Osvaldo, and Gianpiero Fumi, eds. 2006. *Gli agronomi in Lombardia: dalle cattedre ambulanti ad oggi*. Milan: FrancoAngeli.

Federico, Giovanni, and Pablo Martinelli. 2018. Italy to 1938. In *Wine Globalization. A New Comparative History*, ed. Kym Anderson and Vicente Pinilla, 130–152. Cambridge: Cambridge University Press.

Fernández-Prieto, Lourenzo, Juan Pan-Montojo, and Miguel Cabo. 2014. Fascism and Modernity in the European Countryside: A Global View. In *Agriculture in the Age of Fascism: Authoritarian Technocracy and Rural Modernization, 1922–1945*, ed. Lourenzo Fernández-Prieto, Juan Pan-Montojo, and Miguel Cabo, 19–41. Turnhout: Brepols.

Forney, Jérémie. 2011. Idéologie agrarienne et identité professionnelle des agriculteurs: la complexité des images du "paysan suisse". *Yearbook of Socioeconomics in Agriculture* 4: 13–33.

Fumi, Gianpiero. 1999. L'amministrazione dell'agricoltura in Italia negli anni dell'unificazione (1860–1867). In *Temi e questioni di storia economica e sociale in età moderna e contemporanea. Studi in onore di Sergio Zaninelli*, ed. Aldo Carera, Mario Taccolini, and Rosalba Canetta, 379–424. Milan: Vita & Pensiero.

Gabriel (de) Fernández, Narciso. 1963. La agricultura y la escuela en España (1848–1901). *Historia de la educación. Revista interuniversitaria* 2: 131–142.

Gaeta, Davide, and Paola Corsinovi. 2014. *Economics, Governance, and Politics in the Wine Market. European Union Developments*. Cham: Palgrave Macmillan.

Garcia-Parpet, Marie-France. 2009. *Le marché de l'excellence. Les grands crus à l'épreuve de la mondialisation*. Paris: Seuil.

Garelli, Felice. 1873. *Manuale di viticoltura e di vinificazione per gli agricoltori italiani*. Torino: Tipografia C. Favale e Comp.

Giglioli, Italo. 1903. *Malessere agrario ed alimentare in Italia: relazione di un giurato italiano all'Esposizione universale di Parigi, nel 1900, sulle condizioni dell'agricoltura in Italia, in paragone colle condizioni all' estero.* Portici: Premiato stab. tip. vesuviano di E. della Torre.

Girini, Liliana. 2014. *La revolución vitivinícola en Mendoza. 1885–1910. Las transformaciones en el territorio, el paisaje y la arquitectura.* Mendoza: Mendoza Idearium.

Govoni, Paola. 2011. *Un pubblico per la scienza. La divulgazione scientifica nell'Italia in formazione.* Rome: Carocci editore.

Grübler, Arnulf. 2003. *Technology and Global Change.* Cambridge: Cambridge University Press.

Guffanti, Angelo. 1877. *L'avvenire viticolo ed enologico d'Italia, ovverosia la futura principale sorgente di ricchezza nazionale.* Milan: Libreria editrice G. Brigola.

Hames, Gina. 2012. *Alcool in World. History.* London and New York: Routledge.

Harms, Roeland, Joad Raymond, and Jeroen Salman, eds. 2013. *Not Dead Things. The Dissemination of Popular Print in England and Wales, Italy, and the Low Countries, 1500–1820.* Leiden: Brill.

Harvey, Matt, and Vicki Waye, eds. 2014. *Global Wine Regulation.* Pyrmont: Thomson Reuters.

Harvey, Matt, Leanne White, and Warwick Frost, eds. 2014. *Wine and Identity: Branding, Heritage, Terroir.* Oxon: Routledge.

Harwood, Jonathan. 2005. *Technology's Dilemma. Agricultural Colleges between Science and Practice in Germany 1860–1934.* New York: Peter Lang.

Henzell, Ted. 2007. *Australian Agriculture. Its History and Challenges.* Collingwood: Csiro.

Hernández Adell, Ismael, and Josep Pujol-Andreu. 2016. Economic Growth and Biological Innovation. The Development of the European Dairy, 1865–1940. *Rural History. Economy, Society, Culture* 27: 187–212.

Hertzberg, Anna, and Giulio Malorgio. 2008. Wine Demand in Italy: An Analysis of Consumer Preferences. *New Medit* 7: 40–46.

Hurren, Elizabeth T. 2016. The Making of the Modern Land Agent in Victorian Times. In *The Land Agent in Britain: Past, Present and Future*, ed. Beardmore Carol, King Steven, and Monks Geoff, 193–204. Newcastle: Cambridge Scholars Publishing.

Institut International d'Agriculture, ed. 1936. *L'enseignement agricole dans le monde. Europe—2 partie—Agricultural education in the world.* Rome: Impr. de l'Institut international d'agriculture.

Ivone, Diomede. 1982. *Istruzione agraria e lavoro contadino nel riformismo dell'Italia unita (1861–1900).* Napoli: Edizioni Scientifiche Italiane.

Jas, Nathalie. 1997. La promoción de la investigación agronómica en Francia durante el siglo XIX. Louis Grandeau, las estaciones agronómicas y el control de los fertilizantes. *Historia agraria* 13: 195–212.

Jones, Peter M. 2016. *Agricultural Enlightenment: Knowledge, Technology and Nature, 1750–1840*. Oxford: Oxford University Press.

Kookson, Gillian. 2018. *The Age of Machinery. Engineering the Industrial Revolution 1770–1850*. New York: Boydell & Brewer.

Kovatz, Simone. 2013. The Geography of Quality Wine in United Italy Areas and Producers. In *Typicality in History. Tradition, Innovation and Terroir*, ed. Giovanni Ceccarelli, Alberto Grandi, and Stefani Magagnoli, 305–321. Brussels: Peter Lang.

Landon, Stuard, and Constance E. Smith. 1997. The Use of Quality and Reputation Indicators by Consumers: The Case of Bordeaux Wine. *Journal of Consumer Policy* 20: 289–323.

Locatelli, Andrea M., and Paolo Tedeschi. 2015. A New Common Knowledge in Agronomics: The Network of the Agrarian Reviews and Congresses in Europe during the First Half of the 19th Century. In *Une Europe de papier. Projets européens au XIXe siècle*, ed. Sylvie Aprile, Cristina Cassina, Philippe Darriulat, and René Leboutte, 187–203. Villeneuve d'Ascq: Presses universitaires du Septentrion.

———. 2018. Les milieux agronomiques européens et la formation d'un modèle agricole italien au XIXe siècle. *Mélanges de l'Ecole Française—Italie et Méditerranée modernes et contemporaines (MEFRIM)* 130 (2): 299–309.

Lomeni, Ignazio. 1825. *Macchina per la pigiatura delle uve o pigiatore del dottor Ignazio Lomeni*. Milan: Giovanni Silvestri.

Lonni, Ada. 1998. Dall'alterazione all'adulterazione: le sofisticazioni alimentari nella storia industriale. In *Storia d'Italia. Annali 13. L'alimentazione*, ed. Alberto Capatti, Alberto De Bernardi, and Angelo Varni, 567–584. Turin: Einaudi.

Lunardoni, A. 1904. *Vini, uve e legnami nei trattati commerciali*. Rome: coop. Poligrafica Editrice.

Luque Ballesteros, Antonio. 2006. Divulgación agronómica y enseñanza agrícola. Las cátedras de agricultura de los institutos de segunda enseñanza en la década de 1860. *Historia agraria* 38: 119–142.

Macedo, Marta, and Jaume Valentines Álvarez. 2016. Technology and Nation. *Technology and Culture* 57: 989–997.

Maffi, Luciano. 2012. *Natura docens. Vignaioli e sviluppo economico dell'Oltrepò Pavese nel XIX secolo*. Milan: FrancoAngeli.

Maffi, Luciano, Manuel Vaquero Piñeiro, and Paolo Tedeschi. 2020. Notes sur les caves sociales en Italie de la fin du XIXe siècle au fascisme. In *L'utopie au jour le jour. Une histoire des expériences coopératives (XIXe–XXe siècle)*, ed. Alexia Blin, Stéphane Gacon, François Jarrige, and Xavier Vigna, 245–255. Nancy: Arbre Bleu.

Magagnoli, Stefano. 2015. Les produits typiques: une construction urbaine? Réputation et terroir à partir d'une perspective italienne. In *Les produits de terroir. L'empreinte de la ville*, ed. Corinne Marache and Philippe Meyzie. Rennes: Presses Universitaires François Rabelais et Presses Universitaires de Rennes.

MAIC. Ministero di agricoltura, industria, commercio, ed. 1884. *Annali di agricoltura 1884.* Rome: Tipografia Botta.

———. Ministero di agricoltura, industria, commercio, ed. 1894. *Annali di agricoltura 1894.* Rome: G. Bertero.

———. Ministero di agricoltura, industria e commercio—Direzione generale dell'agricoltura, ed. 1900. *Bollettino di notizie agrarie 1900.* Rome: MAIC.

Malatesta, Maria. 2006. *Professionisti e gentiluomoni. Storia delle professioni nell'Europa contemporanea.* Turin: Einaudi.

Manning, Louise, and Jan M. Soon. 2014. Developing Systems to Control Food Adulteration. *Food Policy* 49: 23–32.

Marescalchi, Arturo. 1925. *Le esportazioni del vino italiano.* Casale Monferrato: F.lli Marescalchi.

———. 1926. *La nuova legge sui vini e gli aceti.* Casale Monferrato: F.lli Marescalchi.

Marescalchi, Arturo, and Giovanni Dalmasso. 1979. *Storia della vite e del vino in Italia.* Milan: Editrice Unione italiana vini.

Mazzotti, Omar. 2017. *"Istruire dalla cattedra, istruire coll'esempio!". Conoscenze agrarie e capitale umano in Romegna tra Otto e Novecento.* Bologna: Il Mulino.

Meloni, Giulia, and Johan Swinnen. 2013. The Political Economy of European Wine Regulations. *Journal of Wine Economics* 8 (3): 244–284.

———. 2018. Trade and Terroir. The Political Economy of the World's First Geographical Indications. *Food Policy* 81: 1–20.

Meloni, Giulia, Kym Anderson, Koen Deconinck, and Johann Swinnen. 2019. Wine Regulation. *Applied Economic Perspectives and Policy* 41 (4): 620–649.

Mirri, Mario. 2004. Andare a scuola di agricoltura. In *Agricoltura come manifattura. Istruzione agraria, professionalizzazione e sviluppo agricolo nell'ottocento, II*, ed. Giuliana Biagioli and Rossano Pazzagli, 13–59. Firenze: Olschki.

Mistry, Anisha, and Luca Valente. 2021. The Barolo Appellation of Origin in the Global Market. In *Wine Law and Policy from National Terroirs to a Global Market*, ed. Julien Chaisse, Fernando Dias Simões, and Danny Friedmann, 373–391. Leiden: Brill.

Mocarelli, Luca. 2013. The Long Struggle for the Chianti Denomination. Quality versus Quantity. In *Typicality in History. Tradition, Innovation and Terroir. La typicité dans l'histoire. Tradition, innovation et terroir*, ed. Giovanni Ceccarelli, Alberto Grandi, and Stefano Magagnoli, 323–340. Brussels: Peter Lang.

Mocarelli, Luca, and Manuel Vaquero Piñeiro. 2017. Imprenditoria vinicola nell'Italia centrale: dalla mezzadria ai mercati globali. *Proposte e ricerche* 79: 153–173.

Moe, Espen. 2007. *Global Leadership. The Role of the State in Technological Progress, 1750–2000.* Burlington: Ashgate.
Mokyr, Joel. 1990. *The Lever of Riches. Technological Creativity and Economic Progress.* New York: Oxford University Press.
Moricola, Giuseppe. 1998. Il «commercio insaziabile»: fiscalità e consumi alimentari in età liberale. In *Storia d'Italia. Annali 13. L'alimentazione*, ed. Alberto Capatti, Alberto De Bernardi, and Angelo Varni, 347–378. Torino: Einaudi.
Moser, Peter, and Tony Varley. 2013. The State and Agricultural Modernisation in the Nineteenth and Twentieth Centuries in Europe. In *Integration through Subordination. The Politics of Agricultural Modernisation in Industrial Europe*, ed. Peter Moser and Tony Varley, 13–40. Turnhout: Brepols.
Musella, Luigi. 1992–1993. La politica della «continuità». L'azione del Ministero di Agricoltura in Italia tra il 1878 e il 1897. *Annali dell'Istituto Alcide Cervi* 14–15: 231–240.
Nützenadel, Alexander. 2008. A Green International? Food Markets and Transnational Politics c. 1850–1914. In *Food and Globalization. Consumption, Markets and Politics in the Modern World*, ed. Alexander Nützenadel and Frank Trentmann, 153–172. Oxford and New York: Berg.
Onger, Sergio. 2010. *Verso la modernità, i bresciani e le esposizioni industriali 1800–1815.* Milan: FrancoAngeli.
Ottavi, Ottavio. 1898. *Enologia teorico-pratica. Monografia sui vini da pasto e da commercio e sui vini di lusso.* Casale Monferrato: Tipografia e Litografia Carlo Cassone.
———. 1900. *Enologia teorico-pratica.* Casale Monferrato: Tipografia e litografia Carlo Cassone.
Ottavi, Ottavio, and Arnaldo Strucchi. 1908. *Enologia. Precetti ad uso degli enologi italiani.* Milan: Tecnografica.
Pan-Montojo, Juan. 1995. La administración agraria en España, 1847–1907. *Historia Agraria* 10: 67–88.
Paul, Harry W. 1996. *Science. Vine and Wine in Modern France.* Cambridge: Cambridge University Press.
Pazzagli, Rossano. 2004. Istruzione e nuova agricoltura in Italia: la fortuna del modello di Cosimo Ridolfi. In *Agricoltura come manifattura. Istruzione agraria, professionalizzazione e sviluppo agricolo nell'ottocento, II*, ed. Giuliana Biagioli and Rossano Pazzagli, 255–284. Firenze: Olschki.
———. 2005. From Private Initiative to State Intervention: The Origins of Public Agricultural Education in Italy. In *Land, Shops and Kitchens. Technology and the Food Chain in Twentieth-Century Europe*, ed. Sarasúa Carmen, Scholliers Peter, and Van Molle Leen, 231–246. Turnhout: Brepols.
———. 2008. *Il sapere dell'agricoltura. Istruzione, cultura, economia nell'Italia dell'Ottocento.* Milan: FrancoAngeli.

Perkin, Harold. 1989. *The Rise of Professional Society: England since 1880*. London and New York: Routledge.

———. 1996. *The Third Revolution. Professional Elites in the Modern World*. London and New York: Routledge.

Petrusewicz, Marta. 1991. Agronomia: innovatori agrari nelle periferie europee dell'Ottocento. In *Storia dell'agricoltura italiana, III, Mercati e istituzioni*, ed. Piero Bevilacqua, 295–344. Venezia: Marsilio.

Philippe, Denise, and Sharon Kingsland. 2015. *New Perspectives on the History of Life Sciences and Agriculture*. Cham: Springer International.

Phillips, Rod. 2016. *French Wine: A History*. Oakland: University of California.

Pilcher, Jeffrey M. 2006. *Food in World History*. New York and London: Routledge.

Planas, Jordi. 2016. State Intervention in Wine Markets in the Early 20th Century: Why was it So Different in France and Spain? *Revista de Historia Economica. Journal of Iberian and Latin American Economic History* 35: 175–206.

Prosperi, Adriano. 2019. *Un volgo disperso. Contadini d'Italia nell'Ottocento*. Torino: Einaudi.

Ravizza, Francesco. 1900. *Come si fabbrica il buon vino*. Asti: Scuola tipografica Michelerio.

Ringsberg, Henrik. 2014. Perspectives on Food Traceability: A Systematic Literature Review. *Supply Chain Management: An International Journal* 19: 558–576.

Rodríguez Vázquez, Florencia. 2012. La recepción italiana en la educación agrícola y en la difusión de conocimientos técnicos para la vitivinicoltura de Mendoza, Argentina (1890–1920). *Mediterranea. Ricerche storiche* 26: 539–558.

Rogari, Sandro. 1994. *Proprietà fondiaria e modernizzazione. La Società degli agricoltori italiani (1895–1920)*. Milan: FrancoAngeli.

Rösener, Werner. 1994. *Les paysans dans l'histoire de l'Europe*. Paris: Seuil.

Rossati, Guido. 1900. *Relazione di un viaggio d'istruzione negli Stati Uniti d'America*. Rome: Tip. Nazionale di G. Bertero.

Rossi, Antonio, and Duilio Cortassa. 2021. Exploring Italy's Wine Law Reforms. Experiences, Challenges and Prospects. In *Wine Law and Policy from National Terroirs to a Global Market*, ed. Julien Chaisse, Fernando Dias Simões, and Danny Friedmann, 99–126. Leiden: Brill.

Scarzanella, Eugenia. 1986. Trigo y plate (Grano e soldi): l'emigrazione e l'agricoltura argentina (1870–1914). *Revue Européenne des Migrations Internationales* 2: 91–109.

Segers, Yves, and Roeland Hermans. 2009. Between Ideology and Science: Higher Agricultural Education in Belgium and the Development of a Catholic Agricultural Network, 1850–1914. *The Agricultural History Review* 57: 236–256.

Simpson, James. 2011. *Creating Wine. The Emergence of a World Industry, 1840–1914.* Princeton: Princeton University Press.
Stanziani, Alessandro. 2004. La mesure de la qualità du vin en France 1871–1914. *Food & History* 2: 191–226.
———. 2005. *Histoire de la qualité alimentaire (XIX–XX siècle)*. Paris: Seuil.
Stranieri, Stefanella, and Paolo Tedeschi. 2019. The Role of Quality in Wine Production and Market: European Rules, CAP and New Technology. In *A History of Wine in Europe, 19th to 20th Centuries, Vol. II. Markets, Trade and Regulation of Quality*, ed. Silvia A. Conca Messina, Stephane Le Bras, Paolo Tedeschi, and Manuel Vaquero Piñeiro, 255–274. Cham: Palgrave Macmillan.
Stringher, Vittorio. 1901. *L'istruzione agraria in Italia*. Rome: Unione Cooperativa Editrice.
———. 1911. *L'istruzione e la sperimentazione agraria*. Rome: G. Bertero.
Tarjan, Laslo, Ivana Senk, Srdjan Tegeltija, Stevan Stankovski, and Gordana Ostojic. 2015. A Readability Analysis for QR Code Application in a Traceability System. *Computers and Electronics in Agriculture* 109: 1–11.
Tedeschi, Paolo. 2011. La Politique Agraire Commune ou 'le grand cartel agricole communautaire'. In *Marchés régulés: corporations et cartels*, ed. Margrit Müller, Heinrich R. Schmidt, and Laurent Tissot, 243–259. Zurigo: Chronos Verlag.
Timmons, Todd. 2005. *Science and Technology in 19th Century America*. Westport and London: Greenwood.
Tracy, Michael. 1989. *Government and Agriculture in Western Europe, 1880–1988.* Hertfordshire: Harvester Wheatsheaf.
Trentin, Pompeo. 1895. *Manuale del negoziante di vini italiani nell'Argentina*. Buenos Aires: Tip. Elzeviriana di P. Tonini.
Unioncamere. 2007. *Rapporto sul settore vitivinicolo 2007.* Rome: Retecamere Scrl.
Unwin, Tim. 1991. *Wine and the Vine: An Historical Geography of Viticulture and the Wine Trade*. London and New York, Routledge.
Van Molle, Leen. 2005. Kulturkampf in the Countryside. Agricultural Education, 1800–1940: A Multifaceted Offensive. In *Land, Shops and Kitchens. Technology and the Food Chain in Twentieth-Century Europe*, ed. Sarasúa Carmen, Peter Scholliers, and Leen Van Molle, 139–169. Turnhout: Brepols.
Vaquero Piñeiro, Manuel. 2011. *Da fattori a periti agrari. Formazione professionale e modernizzazione dell'agricoltura in Umbria (1884–1929)*. Perugia: Editoriale Umbra.
———. 2015. Readings for Farmers: Agrarian Almanacs in Italy (Eighteenth–Twentieth Centuries). *The Agricultural History Review* 63: 243–264.
———. 2016. 'Empirici' e 'istruiti'. Fattori e periti agrari in Italia tra XIX e XX secolo. In *Formare alle professioni. I saperi della cascina*, ed. Monica Ferrari, Gianpiero Fumi, and Matteo Morandi, 84–104. Milan: FrancoAngeli.
———. 2019. Italian Wine in Argentina (1861–1914): Market, State Intervention and Wineries. *Storia Economica* 22: 331–360.

Vaquero Piñeiro, Manuel, and Luciano Maffi. 2018. The Diffusion of Italian Wine in the United States (1861–1914). *Rivar. Revista Iberoamericana de viticultura, agroindustria e ruralidad* 15: 176–196.

Vecchi, Giovanni. 2017. *Measuring Wellbeing. A History of Italian Living Standards*. New York: Oxford University Press.

Vivier, Nadine, ed. 2008. *The State and Rural Societies. Policy and Education in Europe, 1750–2000*. Turnhout: Brepols.

———. 2009. Le rôle des élites francaises en faveur du progrès agricole au XIX siècle. Réalités et construction d'une image. In *Élites et progrès agricole: XVI–XX siècle*, ed. Vivier Nadine, 187–205. Rennes: Presses Universitaires de Rennes.

Weber, Eugen Joseph. 1976. *Peasants into Frenchmen: The Modernization of Rural France, 1870–1914*. Stanford: Stanford University Press.

Webster, Sarah. 2007. Estate Improvement and the Professionalisation of Land Agents on the Egremont Estates in Sussex and Yorkshire, 1770–1835. *Rural History* 18: 47–69.

Wilson, Damien, and Sarah Quinton. 2012. Let's Talk about Wine: Does Twitter have Value? *International Journal of Wine Business Research* 24: 271–286.

Wu, Linhai, Xu Lingling, Dian Zhu, and Xiaoli Wang. 2012. Factors Affecting Consumer Willingness to Pay for Certified Traceable Food in Jiangsu Province of China. *Canadian Journal of Agricultural Economics/Revue canadienne d'agroeconomie* 60: 317–333.

Zaninelli, Sergio, ed. 1990. *Le conoscenze agrarie e la loro diffusione in Italia nell'Ottocento*. Turin: Giappichelli.

CHAPTER 3

Winemakers

The diseases that struck the European vines during the second half of the nineteenth century and the simultaneous expansion of international wine markets set the framework within which the first nucleus of Italian modern winemaking entrepreneurship was formed (Charters 2006). In the most advanced wineries, the management of the wine supply chain became more complex and more expensive. The solutions required to combat phylloxera and other diseases (powdery mildew and peronospora) caused a significant increase in production costs: besides, new technologies and business models were necessary. The old cellars had to be replaced by technologically equipped ones which were organised according to the wine production for export. Viticulture progressively became a complex productive sector as an industrial one (Federico 2011). At the beginning of the twentieth century, names such as Pio Cesare, Antinori, Frescobaldi, Martini, Gancia, Ruffino, Laborel-Melini, and many others, including a substantial presence of Southern Italian wineries, gave rise to an emerging winemaking business, thanks to which Italian wine underwent an initial and important phase of internationalisation of their markets (Pinilla and Ayuda 2007). New historic brands and labels were born, and for more than a century, they have remained at the top in world wine market.

With the support of the State, the first Italian wine companies created a real national wine sector overcoming the regional dimension characterising other Italian agri-food industry products such as oil, pasta, or cheese (Swinnen 2010). From a social point of view, the development of the wine

M. Vaquero Piñeiro et al., *A History of Italian Wine*,
https://doi.org/10.1007/978-3-031-06097-7_3

sector favoured the creation of noble and bourgeois entrepreneurship (Colli 2012). In some areas of the country (Tuscany, Piedmont), prevailed the experience of noble families (Tedeschi 2019) finding in the modern wine sector a breeding ground to have new earnings without renouncing privileges and social status as large landowners (Pilbeam 1990; Harsanyi 2005). They invested massively in fixed capital as demonstrated by the extensive investments in machinery, the building of new modern cellars, and the recruitment of specialised workers (Federico 2009). At the same time, the "industrialisation" of the wine production favoured the widespread bourgeois entrepreneurship, which often managed to obtain the *more nobilium* (title, palace) through marriage. So, the development of the Italian wine sector accounted for social as well as economic aspects (Nuvolari et al. 2018).

Together with some private winemakers, from the end of the nineteenth century, the State encouraged the formation of cooperative wineries. The role of cooperatives or social cellars was very important: they stabilised the wine sector by helping to organise the production of thousands of small producers. Their relevance did not decrease in the new millennium: currently, the cooperatives Cantine Riunite & Civ and Caviro are firmly in the lead among companies dedicated to the wine production and marketing in Italy, in terms of turnover and social involvement. The history of the Italian wine sector presents two undisputed traits characterising the dominant entrepreneurial structures: the role of family businesses strongly linked to the territories they belong to (Vrontis et al. 2016) and the solidaristic function of cooperative wineries.

Over the long evolution of Italian winemaking entrepreneurship, some characteristics always remained unchanged. Firstly, the weight of small properties, which was favoured the diffusion of wine cooperatives. When winemaking entrepreneurship was still considered a distant prospect, small farmsteads became the linchpin of market-focused production, even if they continued to maintain connections with their places of origin. The family-oriented aspect was in fact a fundamental trait, and it was connected to the relevance of the artisanal economy for the Italian social and economic development during the twentieth century.

The Italian winemaking entrepreneurship continues today to pass down its history and remind the great transformations taking place in Italian rural society, during the second half of the twentieth century. In many cases, the creation of a winemaking company was the response to the desertion of the countryside by young peasants attracted by the new big cities. Over just a few decades, many farmsteads became dynamic entrepreneurial businesses. It was a change without fracture or overturning: it was the effect of a strong tie with places of origin and the dimensions remained unchanged. There were a few cases of mergers of wine companies, and their aim was to improve the productive systems. In the new millennium, excluding a small number of big companies, the Italian wine sector continues to depend on the strategic choices of numerous family-run businesses having small and medium dimensions.

3.1 Family Companies

The history of the Italian wine sector is an emblematic case-study, broadening knowledge on entrepreneurship in agriculture. By evaluating how Italy became one of the winemaking leading countries around the world—in terms of quantity, quality, and image—it is possible to identify what actors and factors mainly contributed to this result. However, the wine sector is not only highly fragmented but also highly articulated in terms of company size, production types and business models. These different production aggregations express different production types, technological factors, communication strategies, development plans, and relations with customers and stakeholders: this obviously implies that actors and factors are numerous, and it is not so easy to correctly evaluate their actual impact on the performances characterising the Italian wine sector. The ability to build relationships with the right people at the right time, and the creation of networks of relationships at all levels, helped to support this specific business. The information and social dimension are crucial to analyse the solid relationships between wine entrepreneurs who have played a key role in the modernisation of the Italian economy and trade. Entrepreneurs of both aristocratic and bourgeois background contributed to the development of the wine business (Maffi et al. 2021b).

A great contribution to the start up of Italian wine companies arrived from the intervention of the State: it favoured the creation of associations, consortia, and cooperatives (Besana et al. 2017; Chiapparino 1998; Oddy and Drouard 2016; Petrick 2012; Segers et al. 2009), and it also contributed to the strengthening of private initiatives by implementing protectionist policies, signing trade agreements, granting loans and awards, providing machinery and other mechanisms useful for building relationships and market networks. Entrepreneurship is an important mechanism for the economic development and has relevant effects on employment, innovation, and socio-economic welfare (Schumpeter 1934): even if the role of Italian companies and entrepreneurs was obviously fundamental for the Italian economic growth (Amatori and Colli 1999; Colli 2002), it was the strong linkages between these economic actors and public institutions that allowed the improving of the economic sector results and of Italian citizens' quality of life. Public intervention had a key role in the take-off of the national economy (Toniolo 1990, 2013; Giannetti and Vasta 2006; Vasta 1999a, b): this also happened for the Italian wine sector.

The first steps took place in the 1860s, that is immediately after the birth of the Kingdom of Italy (1861): both in Piedmont and in Tuscany few noble families with close ties to France began a phase of experimentation driven by the desire to obtain wines resembling those consumed at the best tables on the other side of the Alps. During the 1840s and 1850s, a real emulation was in fact begun, following the arrival of oenologists who had a good knowledge of the productive techniques used in France (Berta and Mainardi 1997; Maffi 2019). In Piedmont this élite were represented by Marchesa Falletti di Barolo, Count Camillo Benso di Cavour (future Prime Minister of the Kingdom of Italy) and also the royal house of Savoy with Carlo Alberto, King of Sardinia and his son, Vittorio Emanuele II, King of Italy, who were also keen to join the newborn dry Barolo production. The Count of Cavour knew the limits and potential of Italian wine. With regard to luxury wines, he was aware that these types of wine were not only sought after by buyers for their intrinsic value but also by "a certain favour that fashion" gave them. As he thought that the European wealthy classes were accustomed to giving preference on their tables to wines from Bordeaux, Champagne, or the Rhine, he committed himself to the production of Barolo with the aim to realise a real competitor to these wines. He was up to date with the latest European trends, because his travels in France and the United Kingdom had given him the opportunity both to learn about the most up-to-date studies of French ampelography and to "taste" the most fashionable wines. In Tuscany, the breakthrough role depended on a small group of nobles exploiting in a modern way farms they owned (Biagioli 2000). Vittorio degli Albizi, born in France, introduced new winemaking methods in his farms: Pomino, Nipozzano, and Poggio a Remole, which later became part of the Frescobaldi heritage. Another Tuscan nobleman who modified the Italian wine production was the Count Bettino Ricasoli who de facto created the Chianti (Ciuffoletti 2009).

The choices made by these nobles involved in wine production were evidence of a relevant change in landowners' mentality: the possession of land, and in particular vineyards, was no more only a factor of social prestige, it also obliged to develop a culture of enterprise that it was scarcely diffused between landowners, in particular the noble ones. Piedmontese and Tuscan viticulturists wanted to develop a prestigious national wine sector which were able to compete with the French one: this also involved the wines that were served at official banquets of the main diplomatic circles. It was in fact a question of prestige and patriotic feelings, but also of diplomatic balance: the newborn country, backward in industrial development, had to become important for the quality of its rural products.

The effects of this change of mentality were evident in the first post-unification industrial surveys which were made in the early 1870s (*Inchiesta industriale 1874*). The province with the highest number of establishments producing large quantities of wine using modern industrial criteria was Alessandria in Piedmont. Eighteen companies were located in Asti: Francesco Albanese, Argenta & Ottolino, Giovanni Bello, Giovanni Bertoletti, Boano brothers, Antonio Marani, Moriondo & Liprandi, Luigi Pistoni & Co., Prato & Argenta, Rissone & Bonino, Taricco & Co., Giuseppe Vigitello, Solaro & Rustichelli brothers, and Ravizza & Robiolio. They employed about 200 workers during the winemaking period, which lasted about a month: in the other months, their number was reduced to about 50. In Canelli, the Gancia company produced wine and vermouth and employed 34 workers (Cirio 1990). In the same municipality, there were many other wine companies which, although they traded with foreign countries, escaped the statistical surveys. In Acqui, there were two relevant companies: the Beccaro brothers employed 23 workers, while the Menotti employed 6 workers. In Alessandria, Casaletti Giuseppe employed 16 workers. In Strevi operated heirs Braggio, Mantelli brothers, Sebastiano Bruzzono & sons, Giuseppe Reggio, Fezzardi & Pignatelli: in total they employed 20 workers. Other municipalities having important winemakers were Ricaldone, Costigliole d'Asti and Casale Monferrato: in this last Ottavi & Morbelli produced cognac using two 35 horsepower steam boilers and two 24 horsepower steam engines. In the province of Alessandria, there were 33 companies employing 454 workers (397 men and 57 women). On average, these winemakers worked 150 days a year, but there were significant disparities: while in Asti, they worked an average of 30 days a year, in Acqui, Canelli, and Costigliole d'Asti, where production was more diversified, including wine, cognac, vermouth, and champagne, the number of days worked rose to 300, that is they were in activity all the year.

The statistical surveys carried out at the end of the nineteenth century and at the beginning of the twentieth century did not fully reveal the reality existing around a few and isolated large-scale wine production and marketing establishments. In fact, within the much broader sector of "food industries", the new wine establishments were not considered: they remained submerged. In the Italian Statistical Yearbooks, the wine appeared in the section concerning the agriculture: there was an interest in the total quantities of "wine which was made" without any equally significant attention being paid to the different winemaking phases. Besides, it was evident that few modern establishments equipped with modern winemaking plants had no relevance in the Italian labour market. According to the general census on 10 February 1901, the number of "wine technicians and other workers in winemaking plants" was only 457, that is 0.3 per cent of people employed in Italy.

Data were better in the census of factories and industrial enterprises, which were carried out in 1911 (Censimento 1914). There was a considerable change in terms of numbers: on 10 June 1911 there were 1334 "oenological establishments" in the Kingdom of Italy, and this depended on the requisites of the companies involved in bottling wine; they must have the title of "oenological establishment". The Italian oenological panorama was dominated by three regions: Apulia, Sicily, and Piedmont with 651, 168, and 138 establishments, respectively, equivalent to 72 per cent of the Italian ones. Other regions had between 30 and 70 establishments, namely, Lombardy, Veneto, Emilia, Tuscany, Campania, and Calabria. Finally, some regions had fewer than 15 establishments, that is Marche, Umbria, Lazio, Abruzzi, Basilica, and Sardinia.

Data did not change significantly in the 1927 census: 1705 establishments with a total of 8317 employees. However, there were some significant changes in the weight of some regions: Apulia fell to 52 establishments, while Emilia rose to 411; Piedmont strengthened slightly with 184, as well as Lombardy (87) and Sicily (276). Other regions showed a substantial stability. In opposite, a decade later, data showed a strong growth: there existed 8285 wine and vinegar plants, of which 3613 are run by motive power. The number of employees reached 22,000.

The progressive importance that industrial wine production achieved during the Belle Époque was put in evidence by the commercial and industrial trademarks registered by wineries beginning to adopt a precise business strategy, also with regard to the recognition and legal defence of their activities (Nuvolari et al. 2018). In the case of vermouth and liqueur wines, the consolidation of some of the companies that, thanks to the labels used, have deeply marked the history of Italy's image abroad, in particular Cinzano, Martini & Rossi (Carioscia et al. 2012), Fratelli Cora, Campari, and Florio, clearly emerges. The scenario is the same in the field of table and sparkling wines. During the twentieth century, several companies became well-known and economically performant international brands: Gancia, Luigi Bosca, or Emanuele Mirafiori in Piedmont (Cantino et al. 2016); Fratelli Folonari in Brescia; Antinori and Laborel-Melini in Tuscany; Scala and Berner in Naples, and the Duke of Salaparuta in Sicily. If we consider longevity one of the best criteria for assessing a company's growth (Colli 2012; Colli and Larsson 2014), it looks clear that the foundations of positive trends in the wine sector have been laid at the end of the nineteenth century. For patents and trademarks trend, the years before the First World War were in fact crucial and allowed the consolidation of Italian wine entrepreneurship (Suffia, Locatelli, and Besana 2018). The Italian office for patents and trademarks registered 150 wine and liquor files from 1872 to 1910: they became 178 in 1915 confirming the great increasing registered after 1900 (+19 per cent) (Tables 3.1 and 3.2).

Given that the required trademarks cover a wide range of alcoholic products, from wine to liqueurs and spirits, at the end of the nineteenth century, despite being in a backward position, the Italian wine sector was able to grasp the importance of having a solid and recognisable entrepreneurial image. Excluding the production of non-conventional wines and liqueur wines (e.g. vermouth, digestif, or marsala) (Hancock 2009), the wine brands covered two types of activities: trademarks for the manufacture and marketing of wines and trademarks related only to the promotion marketing (Galanti, 1896). Many commercial and productive interests existed around wine and alcoholic beverages in Italy, and they were significantly consolidated during the first 15 years of the twentieth century.

Table 3.1 Chronological distribution of wine brands in Italy (1872–1910)

Years	*Trademarks*
1872–1880	7
1881–1890	13
1891–1900	38
1901–1910	92

Source: Archivio centrale dello stato, Rome, Ufficio italiano brevetti e marchi

Table 3.2 Chronological distribution of wine brands in Italy by beverage type (1872–1915)

Trademarks category	*Number of trademarks*	*%*
Wines	77	45
Vermouth	31	18
Spirit	51	30
Marsala	12	7

Source: Archivio centrale dello stato, Rome, Ufficio italiano brevetti e marchi

Looking at the evolution of Italian wine brands by the example of the most famous wines, Chianti and Barolo, it is possible to note that in the first 30 years of the twentieth century, balance prevailed with relatively small numbers. The 1940s were marked by the war and its effects: there was an awakening that in the case of Chianti coincided with the passage of Tuscany under Allied rule. The wineries in fact proceeded with a general renewal of the brands in 1944. In Piedmont, this change took place after the end of the hostilities, between 1946 and 1947.

The 1950s were dominated by Chianti: this wine represented the first phase of the Italian economic growth. Magazines such as *Le vie d'Italia* (edited by the Italian Touring Club) aimed at reviving tourism and included numerous advertisements of Tuscan wine producers. Many furniture and real estate companies invested in the wine sector because they considered Chianti as a safe and promising investment. In the Piedmont area, this arrived starting with the 1960s, when the powerful qualities of Barolo, "the wine of kings, the king of wines", came to the attention of the Italian public opinion.

Looking at the evolution of the trademarks related to the Antinori and Ricasoli, two of the most well-known and international Italian winemakers (Nuvolari and Vasta 2017), they entered the wine production and marketing sector in the early twentieth century, when they officially registered their first trademarks (Figs. 3.1 and 3.2). During the 1920s and 1930s, they registered a greater number of trademarks especially when the laws for the recognition of typical wines were promulgated, that is in 1924 and 1930. In 1944, they proceeded with a general renewal of the brands: the Marquis Antinori's wine company promoted the renewal of seven brands (Chianti, Orvieto), and Baron Ricasoli's wine company did the same for nine different types of wine (Brolio, Meleto, Montelonti, Arbia). This trend continued reaching a new peak at the beginning of the 1960s, in the perspective of the approval of the law on designations of origin (1962).

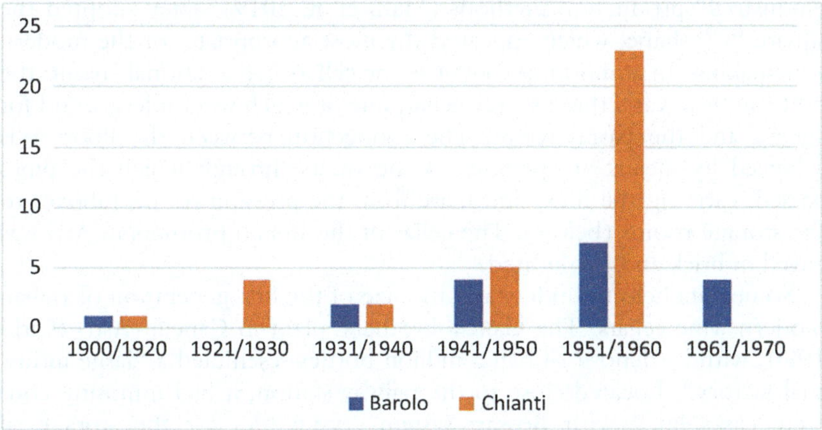

Fig. 3.1 Trademarks Chianti and Barolo (1900–1970). (Source: Archivio centrale dello stato, Rome, Ufficio italiano brevetti e marchi)

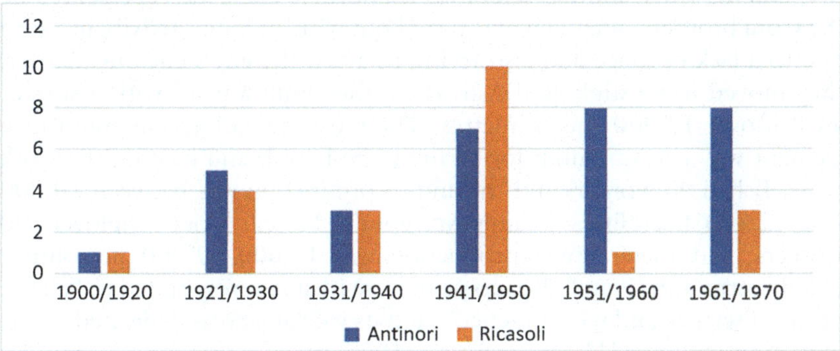

Fig. 3.2 Trademarks Chianti Antinori and Ricasoli (1900–1970). (Source: Archivio centrale dello stato, Rome, Ufficio italiano brevetti e marchi)

At the end of the nineteenth century, in spite of the abundant production of wine, comparing the wine sector with other agri-food ones, there was an increase of new buildings following technical criteria in order to achieve a development in surface area and volume of considerable importance. A new generation of cellars for the production of large quantities of wine put in evidence the transition towards less "artisanal" and more "industrial" production methods (Maffi et al. 2019). They adopted the square "C" shape, which appeared the most appropriate for the modern winemaking. In addition to elegance, the cellars had a rational layout: the best solution was a three-storey cellar, one of which was underground for ageing and the barrel room. The connection between the floors was achieved by means of openings in the vaults through which the pipes passed, carrying the must and marc from the pressing rooms (above) to the storage rooms (below). The cellar of the Boano brothers in Asti was paved in brick and gas-lighted.

Some data help to understand the size of the first generation of Italian modern wine cellars. The Gancia brothers' plant in Canelli (Asti) (Cirio 1990) with its deposit of half a million bottles resembled a "large industrial factory". Located close to the railway station, it had imposing concrete casks lined with Bossari system crystal tiles for the storage of vermouth: their total capacity was 5000 hectolitres. The barrel cellar could hold up to half a million bottles. The annual production amounted to 10,000 hectolitres of different wines and vermouths and 300,000 bottles of sparkling wine. Another relevant Piedmontese plant was represented by the Cora brothers' one. Having started out with an initial activity in Turin due to a lack of space and also trying to avoid the city's customs barrier, they moved to Costigliole d'Asti where they built a plant with a surface area of over 13,000 square metres. The most technologically innovative element was a steam pump for cleaning the barrels and heating the work areas. It had 90 workers and the annual production was between 10,000 and 12,000 hectolitres, mainly vermouth. Fratelli Cora employed 90 workers, and their annual production was 10,000–12,000 hectolitres, mainly vermouth. Fratelli Boano plant located at Asti responded to "all the needs of science and wine practice", and its working areas dedicated to the wine industry looked like a wise fusion of "utility and elegance" (Image 3.1).

Always in Piedmont, the accounting of Luigi Bosca & Figli winery, one of the most active on the North and South America market at the end of the nineteenth century, illustrated the efforts made to know the characteristics of the market demand (Quinn and Oliveira 2019). Carlo Bosca (Luigi Bosca's son) opened a shop in the town of Canelli (Piedmont) for the direct sale of wine and other alcoholic beverages; its network of business relations for the purchase of goods included Tuscany for the supply of Chianti but also Switzerland. Shipments included Italian cities such as Milan, Genoa, Venice, and Parma but also Marseilles in France. It sold Barbera (bitter, amabile) and Moscato spumante in casks; Barbera

Image 3.1 Portrait of Secondo Guasti, Piedmontese winemaker, president of the Italian Vineyard Company in California. (Source: Enotria (1927), n. 10, p. 277)

stravecchio, Moscato appassito and spumante in bottles; Capri bianco and Lacrima Cristi; Chianti (sold in one- and two-litre flasks), Marsala, vermouth and champagne. Chianti was bought for 2 lire and then resold for 2.5 lire: so, it was a 25 per cent mark-up.

The Bosca company, in addition to stimulating the domestic market, was also very active on the international market. To penetrate foreign markets, it took profit by the presence of a large colony of Italian migrants and created solid networks of intermediaries and trusted agents. Personal contacts, place of origin, and reputation resulted to be essential in giving Bosca & Figli the confidence that its strategic business in Argentina was in good hands (Baily 2004; Fauri 2015). Nevertheless, the letters sent by the Latin American country also showed the continuous exchange of news about the development of trade in South America and the evolution of the markets in the United States, starting with New York. Facing the trade globalisation, collecting information useful to direct loads towards the most advantageous business squares became essential for wine companies. To take the right decisions, it was essential to have trusted collaborators and share news between production and trading areas. It was no coincidence that the use of the telegraph in wineries exporting products outside Italy was considered essential. The company's strategy could not do without a network of employees working worldwide and the use of modern communication technologies.

Remaining in Northern Italy, the Maurelio Bassi establishment in Padua, built in 1865 and able to produce from 75,000 to 80,000 hectolitres, was "vast, grandiose, harmonious in all its parts". The plant was built as a working-class village with housing, spacious gardens, and green areas. The Carpenè-Malvolti's plant, built in 1888 in Conegliano Veneto, had branches in Munich and other German cities. The key to its success was the production of luxury sparkling wines, manufactured using innovative machines (De Leo and Favero 2018). Another great plant was designed and directed by engineer Leopoldo Tosi and belonged to the Torlonia princes in Torre San Mauro (Rimini) (Felisini 2019): it produced around 12,400 hectolitres of wine each year. Without ever abandoning banking activities, the Torlonia family invested in the renewal of their new plant: starting in 1875, it was built from scratch to make raw materials handling easier. The cellar could hold up to 50,000 bottles. Prior to the arrival of the engineer Tosi, the revenues from the wine sale reached 15,000 *liras*, and then they increased up to 180.000 *liras*. The main sales markets were Veneto and Lombardy, but also European (Germany, Switzerland), Asian (Dutch Indies), American (Brazil, Argentina), and African (Egypt, Eritrea).

In Tuscany, following the noble entrepreneurship (Kovatz 2013), in the late nineteenth century, the lower-middle-class created "houses" dedicated to the wine sale in Italy and abroad, namely, Caselli and Laborel in Rufina and Luigi Ruffino in Florence. In the 1890s, the sales of Laborel-Melini and Ruffino (Mora 2016, 208–2010), both specialised in Chianti and Tuscan table wines rose considerably. In Italy, Europe, and America, the first increased sales from 267 to 908 hectolitres and the second from 745 to 3904 hectolitres.

In Apulia, the modernisation of the cellars was a direct consequence of the increase in exports of blended wine to France (Dandolo 2010). This is the case of the wineries owned by the French house La Rochefoucauld in the municipality of Cerignola. From 1868, when the vineyards began to be planted, to 1888, the year when the French market was closed, 3500 hectares of vineyards had been planted. Having to store a huge amount of blended wine ranging from 60,000 to 100,000 hectolitres, the construction of eight wineries was carried out.

Another relevant actor was Giovanni Pavoncelli, also owner of a wine plant in Cerignola (Puglia) and a perfect example of a "noble family" of the nineteenth century, enriched thanks to the grain trade. Giuseppe Pavoncelli inherited more than 14,200 hectares (Banti 1996), and his son Nicola (1860–1927) continued in the business world without abandoning the strategy of social strengthening through convenient family ties. In 1883, Nicola married Giulia Monaco of the princes of Arianello. The marriage brought him the chance to have a palace in Naples and a *more nobelium life style*. In the life of the oligarchs of the nineteenth century, honour and social status could and should live with the business. Giuseppe Pavoncelli, thanks to his family ties with the Southern land aristocracy and his role as a Member of Parliament, spurred the winemaking activity and made known his thesis on how to boost the development of national winemaking using the international market. The Pavoncelli firm, in conjunction with the collapse of wine production in France due to phylloxera, expanded the production of blended wine. Between 1877 and 1887, he created a specialised vineyard of more than 2000 hectares. For such a huge amount of wine, he built seven cellars with a maximum capacity of 85,000 hectolitres, which required an investment of more than 600,000 *liras*. Although initially the production was mainly oriented to blended wines, the diversification towards meal wines and special wines soon began. Before 1888, Pavoncelli's wines were sold in the French cities of Nantes, Bordeaux, and Bercy. However, the core business of his activity was not so much selling wines but rather producing them. The closure of the French market, however, forced him to reconvert these strategies. Commercial agencies and warehouses were therefore opened in some main Italian markets (Genova, Milan, Florence, Bologna), with particular care for the consumer cooperatives. On the other hand, the wine production de-localisation abroad proved to be more difficult, despite several attempts in Europe, America, and even Perù that so far had not been included in the Italian trade circuits.

The great novelty of the end of the nineteenth century was the arrival of new entrepreneurship in Southern Italy. In the contrasting modernisation of a sector that involved agriculture in its complexity (Pedrocco 1994), some entrepreneurs undertook a path to innovation, thus giving a boost to change. In Campania, it existed a particular situation: compared to the agricultural sector of Piedmont and Tuscany, where there was an active involvement of agrarian nobility, in Campania the involvement of commercial bourgeoisie was evident (Caglioti 2000). The noble class was still present, but without being dominant (Macry 1988). Local operators mainly worked on the final product processing and marketing: they entrusted the task of growing and ensuring the supply of grapes to inland viticulturists. In this way, agricultural and industrial (i.e. processing) phases were split. This separation was less evident in the noble assets of Tuscany and Piedmont, where all the production phases remained within the same company.

One of the pioneers was Antonio Manzi who founded, in 1832, a company for the marketing of wine in South American countries (Vaquero Piñeiro 2018a, b, 2019). It remained a wine seller company until 1872: then, he started to produce wine in a plant built in Casamicciola d'Ischia. His wine was sold in foreign markets for the first time in 1875, and in 1882, exports reached 42,000 hectolitres considering Bordeaux and Marsala and other wines and liqueurs. However, Antonio Manzi's plant was destroyed by the earthquake that struck the island of Ischia on the night of 29 July 1883: activities restarted four years later with a new plant built in Naples, on the banks of Posillipo hill. The new winery consisted of vast and numerous caves used as a wine cellar for meal wine, dessert wine, bottled, and sparkling wine. The forecourt in front of the plant was occupied by a small garden with vineyards and a series of buildings (houses for the workers, a three-store office building, the owners' family home and alcohol deposits). In the plant overlooking the Gulf of Naples, a small port was built for loading and unloading of grapes and goods. The plant was equipped with a network of tracks that went into the caves making transport easy and cheap. The wine barrels and vats were 108 for a total capacity of 19,208 hectolitres.

In the context of the new-born wine industry in the Gulf of Naples, it was also important the wine company founded in 1852 by the Swiss entrepreneur Amedeo Berner (Canciullo 2013; Hertner 1984). In 1885, his son, Alfredo, moved the company's headquarters from Naples to Casalnuovo, where he built from scratch a winery that required investments for more than 350,000 *liras*: it employed more than 300 people. In 1894, the trademark "A. Berner wine plant in Casalnuovo di Napoli and

Cotonificio Berner in Piedimonte d'Alife" was deposited. It had an unbridled horse in the middle, in a clear reference to the emblem of ancient Greek Naples. The wine of the Berner wineries marketed with the name "Cavallo" (literally "horse") was particularly appreciated in the America markets. Another relevant entrepreneur was Giuseppe Scala di Resina from Ercolano: together with The Laborel & Melini of Florence, Cinzano Francesco in Turin, and Cirio & C. in Turin, he created a company for the opening in London of "a vast emporium for the sale of national wines and food preserves" (Carlucci 1889). Other relevant wineries were those of Camillo Dietz, J. Rouff, and the Marquis Camillo Imperiali in Mercato San Severino (Salerno).

In Sicily, the cellars of Salvatore de Salvo (Riposto, Catania), Corvo (Palermo), and Zucco (Palermo) were noteworthy. The first was half trapezoid in shape, covering an area of approximately 4600 square metres and divided into six areas: the table wine cellar, the dessert wine cellar, the distillery, the barrel factory, and the offices. It produced 40,000 hectolitres, mainly sparkling wines (Moscato and Champagne), and sales concerned the Italian market as well as India, China and Scandinavia. The second was attached to the castle of Duke Salaparuta: the production ranged from 6000 to 10,000 hectolitres. The main export markets were France, Germany, and Switzerland, but there was also positive feedback from other non-European countries such as Egypt, where the relevant (in number) Sicilian community contributed to increasing the demand for Sicilian wines. The Corvo wine was marketed under the deposited (i.e. property right) trademark "Enrico Alliata di Salaparuta". The third was in Terrasini and belonged to Duke Henry d'Aumale. It employed 500 peasants, as the vineyards were all managed by the owner, reaching an extension of 500 hectares. The new plant stood by the sea, and it had a rectangular floor plan with one side 125 m long and the other 50 m long. It was divided into three large warehouses and two smaller ones: they included the barrel cellar, storage warehouses, office, and laboratory. The main product was the "zucco", a liqueur-like dessert wine particularly popular in England and other transalpine countries. In Sicily, it also existed the English merchant John Woodhouse's establishment in Marsala, that is the plant of the "discoverer" of Marsala wine in the eighteenth century. Besides, another important establishment belonged to Baron Antonio Spitaleri in the castle of Solicchiata in Adernò. Its annual production was between 400 and 500 hectolitres of red table wine, known as "Etna Nero". It also produced champagne and cognac (Vaquero Piñeiro 2018b).

At the end of the nineteenth century, in the Italian wine sector an entrepreneurial milieu for international trade was formed (Unwin 1996). This sector proved particularly efficient in addressing the widening of commercial and production horizons. Some of the most important Italian wine companies (Barrio 2008; Cinotto 2012, 2013) were born at the end of the nineteenth century and since then have retained a strong vocation to international trade. Italy participated in the formation of international wine entrepreneurship. The European wine industry was characterised by considerable specialisation, without having the characteristics of a vertically integrated company structure, as it happened in the emerging overseas countries (Argentina, Australia, and California) (Simpson 2011). In the non-European oenology, the main challenge was to create a new sector capable of becoming competitive at a global level (Regan-Lefebrvre 2017), in Europe, as the Italian case shows, the main problem was to modify a consolidated production system. Introducing elements of novelty into a solid tradition dominated by old schemes was not an easy operation. It involved overcoming resistance and opening up to new business and commercial models. This change began from the top, driven by a small group of entrepreneurs who were ready to the competition in the international market.

Wine entrepreneurs followed different strategies to penetrate highly competitive markets such as Germany and the United States (Federico 2004). To sell the best Italian wine abroad, it was necessary to overcome lots of negative conditions (Moser and Varley 2013; Swinnen 2010, 2018). French oenology was a model to follow and emulate (Garcia-Parpet 2009). It was necessary, for example, to learn how to produce wines which were adapted to the foreign consumers' tastes. At the same time, selling abroad meant regulating exports and defending the most renewed wines such as Chianti. It was essential to use advertising and labels on bottles where, in addition to the name of the easily identifiable wine, precise indications on the characteristics of the content had to appear. In particular the laws of the United States, on the wave of protectionism and the restriction of alcoholic beverages (Food and Drugs Act), were particularly demanding in terms of purity and honesty of food products (Law and Libecap 2006). The novelties to be undertaken appeared rich and diversified, contributing to raise awareness of the improvements to be introduced. When the first nucleus of wine entrepreneurship was formed in Italy (Corsi, Mazzarino, and Pomarici 2019), the same happened in other countries, European and non-European. The formation of international wine entrepreneurship in fact responded to common and various commercial agronomic reasons such as the changes in world trade because of the new quick boats and trains as well as the crisis triggered by the diffusion of the phylloxera (Federico and Martinelli 2018).

In the last decades of the nineteenth century, a virtuous circle was progressively created between entrepreneurs and the State (Planas 2016a) that led to the reduction of blended wines sales and the increase of exports of quality wines in bottles. It was the result of public policies oriented to encourage the foundation of winemaking schools, the creation of public stores for winemaking machinery, the organisation of international exhibitions, the creation of cooperatives and the opening, in major European and American cities, of cellar-stores and "oenotechnical agencies" to promote the internationalisation of Italian wines (Pinilla and Ayuda 2002). The main driver was the synergistic relationship between entrepreneurs, associations and formal institutions as confirmed by the role of the General Society of Italian Winegrowers. Institutional interventions were oriented to raise socio-economic development and well-being. Policymakers and practitioners worked to reduce market uncertainties, to minimise transaction costs, and to establish a solid network of social relations (North 1990). Clear rules about cellar-stores guidelines, trade policies, technicians' training, or the legislation concerning hygiene were established. These aspects, hence, allow to compare the Italian case to other countries (Bell 1993; Guy 1997; Heintz 1995; Lapsley 1996; Loubère 1978, 1990; Pan-Montojo 1994, 2001; Paul 1996; Wolikow 2019) where a clear link between wine entrepreneurship training, industrial policy, and innovation was detected (Breckenridge and Taplin 2005).

At the end of the 1880s, the negative consequences of the tariff war against France, which was the main foreign market of Italian blended wines (Anderson et al. 2017), resulted in the public institutions' intervention meant to create the proper conditions for close interaction between private and public initiatives (Fenoaltea 1982, 2011). This collaboration was not limited to the transmission of knowledge and information concerning new modern production systems (vineyard crops and winemaking) and the control of the final product (to improve its quality and avoid adulteration), but it also promoted the establishment of new institutions and the implementation, in Italy, of modern entrepreneurship and business models oriented towards international wine production and marketing (Maffi et al. 2021a).

It is also possible to analyse new wine entrepreneurs observing the ministerial recruitment competition banned in 1885 (Carlucci 1889). Its aim was to reward the production of table wine and the committee accepted 9 out of the 24 applications submitted: Duke Alliata di Salaparuta (Casteldaccia, Palermo), Salvatore De Salvo e figli (Riposto, Catania), Manzi Antonio (Naples), Raffaele Caselli (Rufina, Florence), Luigi Laborel-Melini (Pontessieve, Florence), Filippo Liccidi (Rufina, Florence), Ilario Luigi Ruffino (Florence), Carlo Angelini (Valenza Po, Alessandria), and Count Emanuele Alberto of Mirafiore (Fontanafredda, Serralunga d'Alba). However, none of the participants, who visited between 1886 and 1887, was deemed eligible to receive the award: the production of table wine was in fact less than the 5000 hectolitres, that is the limit required to enter the competition, or there was an evident predominance of blended wine. However, this competition was important: it emerged that, notwithstanding the "agrarian crisis" (Frascani 2012), the wine sector was booming.

At the end of the nineteenth century, thanks to the strategies carried out by the best wine entrepreneurs, best Italian wines were internationally known. The international exhibition of San Francesco (1893–1894) was attended by the Torlonia winery in San Mauro di Romagna, Laborel-Melini in Florence, Ricasoli in Florence, Mirafiore in Fontanafredda, and the Gancia brothers in Canelli. This participation showed the great changes concerning the best winemakers: Italian bottled wines, with labels, reached a great and expanding market where the competition was high. These winemakers also encouraged innovation, change and a comparative approach (Teughels and Scholliers 2017), in a context where "local" agri-food products progressively became "global" and "typical" (Pilcher 2017). The overseas exhibitions were also of particular importance, as they were an opportunity to experiment with the ability to conserve wine on long journeys, and overcome its perishable nature, an essential requirement to enter previously untried new markets. This was the context of The Centennial International Exhibition in Philadelphia in 1876: it was attended by 148 Italian wine producers, with a predominance of Sicilians and Tuscans. The Chicago World's Fair in 1893 hosted 65 exhibitors of Italian wines, which earned particularly positive comments: Italian wine producers exhibited all their wines, including the heavy bodies, still and sparkling ones of the northern mountains, the clear, well-tempered Chianti of Tuscany, the rougher clarets of the Apennines, Marsala, and other Sicilian wines, and the best choices of Vesuvius and the

Neapolitan islands. Another opportunity for a comparison between Italian winemaking and the American context was represented by the Louisiana Purchase Exposition in 1904 (*Official Catalogue of Exhibitors* 1904, 247–250). The 58 Italian wine exhibitors in Saint Louis were from Puglia (17), Tuscany (11) and Sicily (10), then Piedmont (5), Campania (5), Calabria (2), Abruzzo (2), Molise (1), Lazio (1): four wineries had an unrecorded origin (Vaquero Piñeiro and Maffi 2018).

In the last decades of the nineteenth century and in the early twentieth century, in Italy, the wine entrepreneurship began to develop and became crucial, as it still is, for the economic growth of the Peninsula. In this first stage of development, the role of public institutions and private initiatives had great importance. Innovation, investment, skills, and international attitude helped to increase the reputation of Italian wines in the world markets. Far from being a smooth process, the initiatives arose as a response to negative factors, such as the agrarian crisis, the spread of phylloxera, the international economic crisis of the early 1880s. Nevertheless, the institutions and entrepreneurs were able to give rise to a series of activities that generated good results in the long term. As we have seen, not all public and private initiatives led to the expected results, as it occurred for wine societies or the General Society of Italian winegrowers. In general, however, the entrepreneurs, thanks to companies, "oenotechnical agencies", commercial agencies, and cellar-stores in the most important cities of the world entered a circuit of relations that favoured, thanks to institutional intervention, the spread of Italian wine, winning spear of the Italian food. As for historiography, the internationalisation of Italian wine is related to the phenomenon of migration and the purchase of wine by large colonies of compatriots outside Italy. This aspect is not negligible, the cost of Italian wine bottles, however, was certainly not within the reach of all Italian emigrants. The increasing trend of Italian wine, for example in the German area, shows that the internationalisation of the agri-food products of the peninsula does not depend solely on the migration phenomenon. Evidence, as the United States and Argentina show, lends itself to different readings, but the internationalisation cannot be referred to the weight of ethnic demand only.

From the 1920s onwards, the effects of the First World War and the policies imposed by Fascism did not favour the interests of entrepreneurs who were more open to international markets and investment. Many protagonists of the previous phase did not disappear, but the wine sector decreased during the 1920s and 1930s, and business weakened. Observing the registration of new wine companies, it is possible to note that the most recurrent names are those of a limited number of Tuscan and Piedmontese producers whose most prestigious products were Chianti and Barolo (O'Keefe 2014). After the promulgation of new rules concerning the typical wines in 1930, some winemakers referring to the area of origin of their wines to choose the name of their products, that is "colline senesi": they were few as it happened in Emilia-Romagna for the producers of Sangiovese and Lambrusco (Image 3.2).

Image 3.2 Poster of the 10th National Exhibition of Typical and Fine Wines (Siena 1960). (Source: Archivio Storico Unione Italiana Vini—Milano)

Finally, concerning the production of table wines, there was less presence of southern operators, except for a few winemakers in Brindisi and Barletta who represented the consolidated tradition of wine exporters from Apulia.

During the reconstruction after the Second World War, the Italian wine sector was forced to go through a long transitional phase, recovering with difficulty the positions lost. During the 1950s, it was still maintained in Italian wine-growing circles that the future of the sector depended exclusively on the ability to transform Marsala into true Italian wine (Giacalone Monaco 1958). This showed how Italian winemakers were distant from the dynamics of international markets and new models of consumption. Besides, the absence of rules concerning the denomination of areas of origin favoured the best wineries which were able to emulate the products having the best performances on the international markets: this obviously created great problems for local producers which were not able to improve the quality of their wines. A case in point was the wine tasting organised in 1959 by the Society of New York and Wine and Food at the Tavern on the Green in Central Park. On this occasion, along with wines from California and Burgundy, two Lacryma Christi wines were chosen to represent the Italian wine: even if this wine was normally made in Campania, the organisers chose the wines produced by the Marchesi Antinori of Florence and by the Cora brothers of Turin.

3.2 Cooperatives

Wine cooperatives known as "cantine sociali" appeared and became relevant in Italy at the end of the nineteenth century, in connection with the awareness of the structural weaknesses of the Italian viticulture and oenology: it is important to note that the oenological societies, founded by large landowners, nobles and bourgeoisie in the previous decades, had few members and, above all, none had a cooperative structure. Following the positive example already present in other European countries (Planas 2016b; Simpson 2005) and in particular the good results obtained in Germany (where they had more than 200 members and participated in the renewal of viticulture methods and winemaking techniques), the idea of applying the cooperative model to the wine sector began to spread in the Italian terroirs when the diseases concerning vines and the great agrarian crisis got too much difficult for a single winemaker to continue his activity (Cova 1988; Maffi, Vaquero Piñeiro, and Tedeschi 2019). The wine cooperatives in fact came into being in order to provide an adequate response to attacks from cryptogams (powdery mildew and downy mildew) and above all from phylloxera. These diseases favoured the idea, also supported by public institutions, of coordinated intervention by an association of cooperating winemakers to better manage the costly innovations needed to modernise the viticulture and the winemaking process. The cooperative model allowed to raise the financial resources needed to graft European vitis vinifera onto American vines in order to obtain a normal grape harvest. Besides, it allowed to improve the quality of the wine produced and the related sales. Furthermore, it was also the result of some winegrowers' desire to promote an economic and social growth based on solidarity and to become actual players in the development of wine production and trade. The wine cooperatives also enabled their members to free themselves from intermediaries in the sale of their grapes and to less depend on fluctuations in market prices: they in fact managed more efficiently the overly abundant harvests reducing grape prices and permitted a better valuation of the grapes, taking into account production costs. Moreover, they favoured the production of more and better quality wine with facilities (larger, cleaner, and more airy cellars) and equipment (new presses and other winemaking tools), which individual winemakers could not acquire. Finally, for the wine cooperatives the access to rural credit was easier, allowing them to finance the purchase or rental of more modern equipment and enabling optimal use of both the grapes harvested by members and the residual products of the winemaking process (Mignemi 2017; Boletti 1890; Marchesi 1924).

These advantages explained the gradual growth of Italian wine cooperatives: even if quantitative data vary depending on the source used, and it is difficult to calculate the number of members and to evaluate the real assets they owned, the first ones were founded in the 1880s. Until the early twentieth century, their diffusion was very slow, and their presence was sometimes brief: their structural and financial fragility often hindered the implementation of the innovations concerning the winemaking process. It was only in the 1910s that the wine cooperatives began to play an important economic role and have a significant impact on improving the quality of certain Italian wines.

Initially, public institutions promoted the modernisation of the wine sector by awarding monetary prizes to encourage the creation of wine cooperatives. In October 1881, a public competition was organised to award prizes to the best wineries and, in the following years, other initiatives were aimed at improving the efficiency of winemaking. In 1887, a competition was launched to promote the creation of cooperative wineries, but only two of them participated: the Cantina Sociale Pestellini, the first wine cooperative in Italy, founded in Bagno a Ripoli, near Florence, by the lawyer and oenologist Ippolito Pestellini, a member of the Accademia dei Georgofili (the main Italian institution dedicated to agronomic studies), and the Unione Vinicola Brindisina founded in Brindisi, in Apulia. Three years later, the Ministry financed a new competition for wineries. In the mid-1890s, there were six of them, while there were already 200 social dairies: their success motivated winegrowers in favour of the cooperative model (Lissone 1895).

The spread of phylloxera obliged the public institutions to create consortia against phylloxera (anti-phylloxera consortia) whose aim was to provide winegrowers with American vines to graft onto native vines. Cooperation in the fight against phylloxera also led to cooperation in the later stages of the grape harvest, especially since the results obtained by other agricultural cooperatives provided a solution, saving the wine sector and also developing it. Cooperation reduced the structural weakness of small-scale winegrowers who, despite the most prosperous harvests, were unable, due to a lack of money, to invest, to use more modern winemaking systems and thus to improve the quantity and, above all, the quality of their wines. In addition, the cooperative model got easier to resist the negative effects of the new wine tax of 1897. However, there existed only 30 wine cooperatives at the end of the nineteenth century. The spread of these cooperatives was slowed down by the lack of capital for equipment and the winemaking process: moreover, it was difficult to find managers with the necessary oenological skills, good mechanical knowledge, and experience in wine distribution and sales. Finally, many winegrowers did not want to abandon the old productive methods entrusting their grapes to associations which also included their competitors: they were used to easily selling poor-quality wine and did not understand why they had to invest in a modern winemaking process.

The diffusion of the phylloxera and the consequent drop in income forced winegrowers to graft vines using public consortia: the wines of the best winegrowers (including those of some wine cooperatives) began to be appreciated by consumers, while the sales of other producers declined. Prices of wines were in fact increased and customers paying more asked for more quality wines. Public institutions gave prices for better quality wines with the possibility of accessing the international market. In this context, the idea of joining a cooperative became more attractive and their number increased, as did the quality of their products. To achieve this, public institutions continued to promote the creation of new wine cooperatives: in 1902, a new competition was launched for associations and wineries producing wines of "constant type". Public institutions sought to develop the production of wines with characteristics that were easily recognisable to consumers and that could create a bond with them. It was not a question of denomination of origin, but of giving the wine of each terroir a good quality and the possibility of differentiating itself from others. This goal was evidently easier to reach by winemakers who were associated with a cooperative.

In Piedmont, these cooperatives began to select the vines and grapes best suited to making certain types of wine with a good market value: they then identified the best characteristics of a terroir and used a single type of vine. This terroir sometimes had the size of a single commune, as in the case of Barolo, or of several communes in several departments, as in the case of Barbera. The increase of sales in the national market was the first objective: taking advantage of the demand from immigrants, niches were also being sought in the international market (Nützenadel 2008; Magagnoli 2015).

The creation of new cooperative wineries brought the number of members to 2958: the related assets arrived at almost 160,000 Italian lire with a turnover of 1,500,000. The difficult economic situation between the years 1902 and 1903 and the greater resistance of the cooperatives to the increase in production provoking the fall in grape prices, a phenomenon also seen in other Mediterranean wine regions such as Catalonia (Planas 2016a; Simpson 2000), favoured the creation of new wine cooperatives (sometimes it was the reopening of closed wineries): they tried to rationalise the wine production and distribution and to lobby the government (such as it happened for the Consortium among landowners of Monferrato, the Society of Viticulturists with Wine Exchange of Casale Monferrato, and the Cantine Castelli Alto Monferrato).

Many social wineries were linked to the Catholic movement and financed by Catholic banks. In some cases, they are founded directly by parish priests or pastoral councils (e.g. at Tremosine and Prandaglio in Eastern Lombardy, or at Monforte d'Alba, in Piedmont) and each member had to belong to a Catholic association. There were also cooperative wineries founded by oenologists receiving their graduation in special schools of oenology or founded by landowners. In these cases, they were linked to the Catholic or Socialist movements, but they could also be ideologically neutral.

Most cooperatives had a small capital raised by their members. Other financial resources came from loans granted by local cooperative rural banks (the "Casse Rurali"), but also from greater ones as the "Casse di Risparmio" (savings banks) or the "Banche Poplari" (popular banks). This external financing was fundamental because it allowed payments to winemakers' members to be advanced, especially in the case of cooperative wineries with a limited capital or they had none and used members' buildings and equipment (Coletti 1901).

Some cooperative wineries work with consumer cooperatives. For example, in Mombaruzzo (province of Alessandria in Piedmont), the local wine cooperative worked with the Avanti consumer cooperative, which was responsible for the annual purchase of 4000 litres of wine. In this regard, during the 1890s, there was a debate on the role of consumer cooperatives in improving the quality of Italian wines (L'enotecnico 1895). In addition, consumer cooperatives (just like the commercial companies involved in selling wine) valued collaboration with wine cooperatives, especially if the latter could supply a good quantity of wine with a stable taste. However, they were wary of the first large wineries competing on the wine market (Atti del congresso enologico italiano 1903, 49–50).

The number of associates varied greatly, from less than a dozen to several hundred, as in Montù Beccaria, in Pavia (Lombardy), exceeding 360 associates in 1905 (Maffi 2010). It was possible to distinguish between the founders, also called permanent or effective, and the temporary (or aggregates) who were associated only during the grape harvest and whose presence obviously multiplied the number of people involved. In Alba, in the province of Cuneo (Piedmont), the number of members increased from 26 to 58, while in Valenza, in the province of Alessandria (Piedmont), it rose from 17 to 67. The founders owned the shares (or quotas) of the cooperative and, in some cases, also the buildings for the winemaking. They were also obliged to ensure the storage of a minimum quantity of grapes (e.g. in Stradella, province of Pavia, it had to exceed 100 q.) and a quantity of barrels and jars of wine proportional to the grapes given. Some wine cooperatives set maximum limits: no more than 40 quintals of grapes, which allowed to have small winegrowers among their members who evidently did not want to mix with large landowners. Finally, other cooperative wineries charged a sum of money for every 5 or 10 quintals of grapes brought in: this obviously helped to make their activity fully self-financing.

Most of them were small and sought above all to involve and help small and medium-sized winegrowers: many viticulturists distrusted large capitalist companies and thought that only cooperatives could really involve small producers and enable them to produce better quality wines and, at the same time, to reduce winemaking costs (Giornale vinicolo italiano 1896).

Members also had obligations regarding the quality of their grapes: in Piedmont, they could only bring Nebbiolo used for fine wines such as Barolo or Barbaresco. In San Damiano d'Asti (department of Asti), they also had to indicate the vineyards of origin to allow the supervisory board to control the grapes. The price paid to members depended on the quantity of grapes, but above all on the degree of sugar indicated by the glucometer. In some cases, the winegrowers have to cultivate the vineyard according to the recommendations of oenologists recommended by the cooperative winery. This attention characterised the best cooperative wineries that wanted to realise good-quality wines. Others simply took their members' wine (as it happened, e.g. in some Sicilian cooperative wineries) or did not check that oenologists' advice was followed (for example in the province of Brescia) (Maffi 2010; Tedeschi 2003).

Regarding the payment of grapes, disparities were linked to the size and therefore the capital of the cooperative wineries. The most common solution was to give an immediate advance in proportion to the quantity of grapes brought in: the value of the advance was established in advance by a committee or directly by the members' assembly. In cooperative wineries without capital, the administrators naturally used the credit obtained: the balance (linked to actual wine sales) was then reduced by the administrative costs and bank interests.

Concerning the technology, wine cooperatives used modern machinery, even if they were not always the most advanced existing on the market: the best ones used concrete vats with the inside covered with "crystallised" tiles, which allowed a large quantity of wine to be stored. The most efficient cooperative wineries entrusted the management of the winemaking process to oenologists coming from oenology schools and guaranteeing an improvement in the quality of their wines. When a cooperative winery produced good wines, there was an emulation effect among other wineries having the capital to make the same investments in modern equipment and qualified staff.

In 1904, the Ministry of Agriculture granted subsidies to wine cooperatives as well as associations of wine producers: one million lire was distributed among small winemakers and wine cooperatives to renovate or build new cellars that could use the wine vats and large barrels of wine lent by agricultural machinery depots. The rules laid down by the Ministry defined the concept of small wine producers and, above all, shaped the structure and organisation of the wine cooperatives. Small producers produced no more than 70 hl of wine on their terroir (which they owned or enjoined in usufruct) or no more than 120 hl on land obtained under emphyteusis. To qualify for the subsidy, it was also necessary to indicate the location and surface area of the terroir, the number of vines and the average production over the last three years. This information allowed public institutions to know the actual diffusion of the small-scale viticulture and thus to assess the best measures to take to continue the modernisation of the productive system. The wine cooperatives were at least three years old, and their members had to produce at least 400 q. of grapes per year. In addition, they had to prove that the members' wine was sold by the cooperative winery only. These provisions increased the size and duration of new cooperative wineries: the statutes naturally established a minimum duration of three years or gave no indication, leaving it unspecified (Maffi et al. 2020).

In May 1909, the Società dei Viticoltori (literally the Viticulturists' Society) organised the first national congress of the cooperative wineries in Casale Monferrato, Piedmont. The winegrowers wanted to form federations to increase their influence on the government (or at least on local elected officials), to improve the winemaking process, and to better coordinate sales. The first federation was founded in Alessandria in 1908 (including 14 cooperative wineries), and the second one was founded in Stradella (Pavia) in 1910 (with 3 social wineries including 2000 small winegrowers) (Maffi 2010). However, these experiments failed after a short time, due to the coordination difficulties among the managers of the associated wineries: they in fact felt more like competitors than cooperators. While federations had no good results, the number of wine cooperatives increased as well as the average quality of their wines. On the eve of the Great War, there were almost 150, and the best ones were able to cope with a wine market that was richer but also more competitive and, above all, demanding products with a consistent flavour (Etudes 1914). The conflict halted their development (and sometimes caused their closure) because of the limits placed by the military authorities on the free sale of wine (as well as requisitions for deliveries to the army). Besides, the front line significantly affected the cooperative wineries located in Eastern Lombardy and Veneto.

Being able to sell in a market that had become free again, the number of cooperative wineries increased as soon as the war ended: some wineries that closed during the war were active again and cooperatives in territories that previously belonged to the Habsburg Empire became Italian. There were in fact 39 cooperative wineries in the Tridentine Veneto and 13 in the Julian Veneto. This "entry" strongly increased the number in the northern regions, where there existed 18 cooperative wineries in Emilia-Romagna (especially in the province of Modena), 9 in Lombardy (especially in the department of Pavia), and 5 in Veneto. In Southern Italy, there was only one winery in Apulia, two in Sardinia, and three in Sicily.

After the war, federations were created at departmental level and a national federation to coordinate the activities of the associated wineries and to represent their interest before the Ministry of Agriculture. In 1918, the Associazione Italiana delle Cantine Sociali (literally the Italian Association of Social Wineries) was founded in Novara (Piedmont), while the federations of the departments of Modena, Reggio Emilia, and Pavia were set up in 1919. On 20 March 1922 the Federazione Nazionale delle Cantine Sociali (literally National Federation of Social Wineries) was created with its headquarters in Modena and also included the wine cooperatives of Piedmont and Sicily. It organised congresses (e.g. in Vercelli in Piedmont in October 1924) and conferences to highlight the importance of the wine cooperatives for the development of the Italian oenology, as well as meetings and training courses for the oenologists in charge of managing them. It also prepared a model statute and a sample constitution for future cooperatives: it remained active until 1931, before being absorbed into the corporate structures of the Fascist regime.

In 1924, the number of members was about 6000 but the size of the wine cooperatives significantly varied. The greatest one was in Marsala (Sicily) and had 500 members, while the others had much fewer: a quarter of them in fact had between 100 and 250 members, and all the remaining ones had only a few dozen. This disparity naturally led to great differences in production and equipment: official data indicated an average annual production of 8000 hl, but in Marsala the production was 30,000 hl, while in Lombardy and Piedmont the production was almost 5000 hl on average. Furthermore, in Emilia-Romagna, the average production was higher (10,000 hl), thanks to the wine cooperatives in Formigine and Nonantola (Modena) reaching up 20,000–25,000 hl. Finally, in the former Hapsburg territories, the annual production was between 5000 and 10,000 hl.

Under the Fascist regime, the number of wine cooperatives increased: in 1928, they were 84, and they became 115 in 1931 (*La cooperation agricole* 1931). They had 10,732 members and produced one million hectoliters of wine: the *Federazione Nazionale delle Cantine Sociali* created in Trieste a general warehouse. In order to increase the construction of cooperative wineries and oenopolies, the law of 29 December 1930, n. 1737, recognised these establishments as being of public utility because they ensured the conservation of products and rationally regulate trade. In 1938, there were 184 cellars with a total of 18,994 members and an invested capital of 5.7 million lire; the storage capacity had increased to 4.8 million litres of wine, and the average production was 1.3 million hl (Statistical Yearbook of Italian Agriculture 1936–1938). Wine cooperatives were concentrated in Northern Italy; there were 127, while in the remaining regions there were just 36. The Emilia-Romagna was the most relevant region with 47, but in terms of capital the most important was the Venezia Tridentina with 2 million lire. In addition, there were also the cooperative distilleries, which in 1938 were 14, with 4997 members and an actual paid-up capital of 2.7 million lire. It was precisely the regime's lack of support for the wine sector (as the regime wanted to produce more grapes for the table and did not like wine which did not help to train good soldiers) that increased the winegrowers' attitude to form cooperative organisations: these latter in fact helped to reduce production and sales costs, and improved the quality of the wine produced. They benefitted their members adopting new winemaking technologies and giving advice on the work to be done in the vineyards to obtain better quality grapes. This, in addition to the improvement in quality and quantity of Italian agrarian schools, allowed to produce in the 1930s a wine whose average quality was much better than at the end of the nineteenth century: only a part of the wine cooperatives was able to produce really high-quality wines because investments in new technologies were still limited, but the difference with the wine realised during the Belle Époque was evident.

After the Second World War and during the economic boom, wine cooperatives represented for many viticulturists the only way to achieve a real modernisation of their production (Casalini 1953). They grew up in number and dimension also because political élites believed that they were the right instrument to definitively develop the Italian wine sector. Among the main supporters of the cooperative wineries were the Ispettorati provinciali di agricoltura, that is the main public institutions at provincial level. The wine cooperatives in fact allowed to introduce relevant changes in the way vines were raised and exploited: the aim was to avoid the dangerous return of phylloxera attacks and, at the same time, to favour the mechanisation of the winemaking. This was fundamental in a context characterised by changes in the wine consumption which made it urgent to reduce production costs and increase the organisational efficiency of the wine companies (Betri 1988). The old cellars were outdated and had to be renewed: this was expensive and it was difficult to realise because of the progressive fragmentation of many large properties creating a widespread network of micro-farms managed by small landowners or sharecroppers: they have no capitals, no suitable equipment, and also no qualified staff. So, associated grape processing granted by wine cooperatives was considered a genuine step forward: Veneto, Emilia and Piedmont were examples of modern and efficient cooperative wineries.

The available data on the evolution of cooperative wineries in Italy indicated a robust growth: they rose from 169 in 1951 to 264 in 1952 with a processing capacity of 350,000 tonnes of grapes equivalent to 3.8 million hectolitres of wine. At the beginning of the 1960s, the number continued to grow, and they became 606. They were also favoured and strengthened by a special and vain fiscal regime that was strongly criticised: according to many opinionists, they were not to dedicate themselves to the sale of wine but only to produce wine and then transfer it to merchants. The storage capacity of wine cooperatives reached 22.5 million hectolitres, almost a third of the national wine production (Botteon et al. 1968). Puglia (131 wine cooperatives) and Emilia-Romagna (104) predominated, followed by Piedmont (86) and Veneto (54).

The effects of the second Italian green plan approved in 1966 to stimulate the development of national agriculture were important: in particular, the Ministry of Agriculture granted financial and technical aid to rural cooperatives. At the same time, the green plan aimed at stabilising prices and organising agricultural markets by supporting the harvesting, preserving, processing, transforming, and selling of products. As the main problem for wine cooperatives was the lack of financing and bank credit, the government promoted public investment, especially in the southern regions. So, the Cassa per il Mezzogiorno (Dell'Osa 2019) dealt with this and invested to promote the agricultural modernisation in Italian backward areas (Moser and Varley 2013): the wine cooperatives financed by the public institutions increased and, normally formed by small viticulturists, helped to solve main problems of the local winemaking, but also acted as a transmission belt between local political power and the social and economic interests revolving around the wine sector.

In the 1970s, the number of wine cooperatives grew up: they were 619 and had 254,520 members. They showed a high production capacity: grapes processed rose to 27.3 million quintals, wine produced to 16.4 million hectolitres and the possibility of storage to 30 million hectolitres. They put different types of wine on the market: blended wine, 2.3 million hectolitres; ordinary wines or wines for current consumption, 14.7 million hectolitres; wines with denomination of origins, 1.6 million hectolitres (*La cooperazione vinicola italiana*, 1974: 13). Even if the law on designation wines entered into force in 1963, the quantities in the upper wine segment remained still relatively small compared to bulk table wines. In 1972, a new law concerning the wine cooperatives was promulgated with the aim to reorganise the wine sector which had been growing for 20 years without a precise regulatory framework: it proposed tax advantages favouring people carrying out rural activities.

The increase in number continued until 1985 when they reached a peak of 868: then they progressively fell to 677 and at the mid-1990s further reduced to 91 with 51,545 members and a turnover of 988,073 million lire.

3.3 New Trends

According to the sixth industrial and commercial census carried out in Italy in 1971, at the beginning of the 1970s there were 10,324 active wine companies in Italy: 156 produced special and sparkling wines. Individual enterprises predominated on companies and cooperatives. The situation progressively changed, and in recent years, there was a significant expansion of the leading companies: they in fact increased their production-distribution network as well as their shareholdings in other wine companies. In the early 2000s, there was an intense process of company acquisition and concentration involving medium-large companies that were participated in or acquired by the leading ones. These latter extended their presence by increasing the number of local units, that is the establishments producing wine: between 2000 and June 2006, they more than doubled (from 67 to 136), and the greatest increase was recorded in Southern Italy, where 35 new openings were offset by only 11 closures (Unioncamere 2007).

Turning now to look at the organisational-strategic aspects and commercial performance of the leading companies, it is possible to note that about 70 per cent grew up in their turnover in the three-year period 2003–2005.

The 2019 turnover of the top 30 wine companies active in Italy, 4.645 million euros, confirmed the weight of the three most important ones: Gruppo Cantine Riunite & CIV (612 million), Caviro (330 million), and Palazzo Antinori (234 million) (Corsi et al. 2019). Already in the mid-1990s, GIV (Gruppo Italiano Vini which entered Cantine Riunite & CIV) and Caviro were among the leading Italian groups in the beverage sector, even if they were far from the top spots occupied by the Peroni (Beer) and Cinzano groups (Table 3.3). Compared to this situation, the situation before the Covid pandemic appeared better for the wine companies. The first group resulted from the union of the GIV with the Cantine Riunite & Civ. GIV was a wine cooperative in Verona consisting of eight associated cooperatives and operating in the winegrowing, carrying out cultivation and grape crushing activities. It associated 71 wineries throughout Italy and managed 1200 specialised hectares, located in the most important winegrowing regions, in particular: Piedmont (Monferrato), Lombardy (Valtellina Superiore), Veneto (Bardolino and Valpolicella), Tuscany (Chianti Classico), Umbria (Orvieto), Lazio (Frascati), Sicily (Alcamo), Basilicata (Vulture), and Apulia (Salice Salentino). GIV invested abroad not in wine production companies but in distribution or import companies of Italian products: the American Frederick Wildman & Son Ltd, the French-Belgian Carniato and the Canadian Peter Mielzynski. Cantine Riunite & CIV was created in 1950 from the union of nine cooperative wineries in the province of Reggio Emilia: they were later joined by another group of cooperative wineries located in the province of Modena, the Consorzio Interprovinciale Vini (CIV). The group consisted of 1700 members working 4000 hectares of vineyards and enjoyed a solid position in the sparkling wine segment (Lambrusco, Prosecco).

The second wine cooperative, Caviro, was also founded in the Emilia-Romagna, in 1966. It currently unites 30 cooperatives, located in seven regions (Veneto, Emilia-Romagna, Tuscany, Marche, Abruzzo, Sicily, and Apulia) for a total of 36,000 hectares of vineyards. So, compared to a wine supply chain dominated by fragmentation into many small companies, many of them producing high-quality wines, these two large wine cooperatives realised a relevant process of concentration.

They were followed by other wine companies: Casa Vinicola Botter (202 million), Fratelli Martini (214 million), Casa Vinicola Zonin (202 million), Cavit (190 million), Mezzacorona (188 million), Enoitalia (182

Table 3.3 Top 30 wine companies in Italy (2019)

	Region	*Revenue*	*Production*	*Type*
		(million euro)	*(millions of bottles)*	
Cantine Riunite	Emilia-Romegna	612	211	Cooperative
Caviro	Emilia-Romegna	330	71	Cooperative
Palazzo Antinori	Tuscany	234	25	Family
Fratelli Martini	Piedmont	214	**	Family
Casa Vinicola Zonin	Veneto	202	51	Familiare
Casa Vinicola Botter	Veneto	195	97	Mixed
Cavit Cantina Viticoltori	Trentino	190	**	Cooperative
Mezzacorona	Trentino	188	50	Cooperative
EnoItalia	Veneto	182	109	Family
Santa Margherita	Veneto	177	23	Family
Italian Wine Brands	Lombardy	150	63	Mixed
Cantina sociale Soave	Veneto	141	35	Cooperative
La Marca vini e spumanti	Veneto	140	57	Cooperative
Terre Cevico	Emilia-Romegna	131	83	Cooperative
Collis Veneto Wine Group	Veneto	119	25	Cooperative
Compagnia Frescobaldi	Tuscany	115	12	Family
Ruffino	Tuscany	110	34	Foreign
Campari	Lombardy	110	**	Family
Mondodelvino	Emilia-Romegna	107	56	Mixed
Schenk Italia	Bolzano	103	60	Foreign
Vivo cantine	Veneto	102	31	Cooperative
Lunelli	Trentino	101	11	Family
Contri Spumanti	Veneto	95	72	Foreign
Villa Sandi	Veneto	94	26	Foreign
Mionetto	Veneto	86	**	Family
Vinorte	Apulia	86	**	Cooperative
Cantine Ermes	Sicily	85	**	Mixed
Farnese	Abruzzo	84	25	Family
Vinicola Serena	Veneto	82	**	Cooperative
Cantina di Conegliano	Veneto	80	**	Cooperative

Source: *Indagine sul settore vinicolo* (2020)

million), and Santa Margherita (177 million). Casa Vinicola Botter is the export champion in 2019 with 93.7 per cent of turnover, followed by Farnese (92 per cent), Ruffino (91.4 per cent), Martini brothers (86.1 per cent), Mondodelvino (83.3 per cent), and La Marca (82.8 per cent). Wine cooperatives and family groups were in almost perfect parity in terms of the total amount of turnover and the number of bottles, although at the top of the list there were the cooperatives Cantine Riunite & Civ and Caviro.

In 2018, the turnover of wine cooperatives reached a record 5.2 billion euros, accounting for 40 per cent of the Italian wine turnover. The growth was +9.2 per cent in 2017, driven by sales on the domestic market (+13.6 per cent). More than a third of turnover was generated in foreign countries: in 2018, the export of cooperative wineries reached the record figure of 2 billion euros, a share equal to 31 per cent of the total value of Italian wine exports, which reached 6.2 billion euros in 2018. More than 9000 workers in 480 cooperative wineries (two-thirds employed on a permanent basis), 141,000 adhering members, and production equal to 58 per cent of Italian wine described the relevance of the Italian wine cooperatives. The value of members was guaranteed by an average level of mutualist prevalence of well over 82 per cent.

Data on wine production in 2016–2017 indicated that wine cooperatives in France, Spain, and Italy produced 75 million hectolitres, accounting for 45 per cent of EU wine and almost 30 per cent of world production. In France, 65 per cent of winegrowers were members of a cooperative winery, and the 650 cooperative wineries accounted for 48 per cent of French production. In Spain, there were 448 cooperative wineries accounting for 60 per cent of national wine production.

Cooperatives accounted for half of Italian PDO production and 65 per cent of PGI production. In Veneto, 80 per cent of Soave DOC, 62 per cent of Valpolicella, 53 per cent of Custoza, and 50 per cent of Prosecco came from cooperative wineries. In Trentino, the wine cooperatives reached percentages of over 90 per cent for Teroldego Rotaliano, Trentino, Valdadige, and Casteller DOC wines. Wine cooperatives in Emilia-Romagna had the primacy with over 90 per cent of the various Lambrusco denominations and 75 per cent of Sangiovese di Romagna. In Tuscany, they produced 50 per cent of Nobile di Montepulciano, while in Piedmont 42 per cent of Dogliani is cooperative. In Apulia, wine cooperatives produced more than 80 per cent for six PDOs.

In 2018, the supply of wine remained rather fragmented: there were only 155 companies with a turnover of more than 25 million euros for a total turnover of 7.2 billion euros. The geographical distribution of the 30 top wine companies showed the primacy to the Veneto with 1.695 million euros (about 36.5 per cent); Emilia-Romagna followed with 1.180 million euros (25 per cent). In these regions the presence of cooperatives was very strong. A similar situation arrived in Trentino where the three top companies, two cooperatives and one belonging to a family group (Lunelli), had a turnover of 479 million (10 per cent). Tuscany, with 459 million euros (almost 10 per cent), was mainly represented by family groups (Palazzo Antinori, Frescobaldi, Ruffino). Far behind were Lombardy with Campari and Piedmont with Martini brothers, while other regions had a little weight. So, there existed a concentration of large wine companies in only two regions (Veneto and Emilia-Romagna): this expressed the strength achieved by these territories in the new millennium. At the same time, Apulia, representing 18 per cent of the Italian produced wine, remain out of the ranking, demonstrating the deep-rooted weakness of the southern wine companies which are still strongly linked to the production of large quantities of wine.

In general, family companies grew up following the evolution of the families themselves, but there were also some cases of union of wine producers. Sella & Mosca was bought by Campari in 2002, and the Antinori group incorporated other groups. There were no real leaders and foreign capital was rare: at the same time, there was a lack of Italian investment abroad.

All actors operating in the Italian wine sector participated in the process of diversification of production in the wine sector in Italy: in recent years, this sector in fact opened up to organic, bio-dynamic wines and other forms of cultivation of vines that refer to the problems of the sustainable development model. From this point of view, wine is part of an overall cultural reform of the primary sector that is much more attentive to the principles of safeguarding and enhancing biodiversity and respecting the environment and consumer health. This evolutionary process is still in full swing and confirms the continuous evolution of the winegrowers who adapt their terroirs to the demands of society and institutions. The pace of adaptation to the processes underway is not the same, and there are producers who show less capacity to respond with consistency to the market changes (Mora 2006). Where the process of adaptation occurred with greater speed, a process was consolidated and marked the emergence of entrepreneurs who wanted to participate in the construction of a different way of practising and understanding the agrarian activity, including the viticulture (Ramón Resina and Viestenz 2012).

The so-called more radical theses of neo-ruralism or neo-conservatism lead to calls for a return to the countryside as a way of recovering an ancestral way of conceiving the relationship with the land and the practice of agriculture. Research in the demo-anthropological field conducted on this specific area of business (Le Guénanff 2020) reveals a storytelling from which the profile of a generation of entrepreneurs emerges, young and educated, who declare to defend a model of ecological agriculture antithetical to the logic of the market and large speculative profits. In 2008 (Gilinsky et al. 2016; Santini et al. 2013) the Federazione Italiana Vignaioli Indipendenti (FIVI) was founded in support of environmental sustainability in the wine sector, as part of the European Confederation of Independent Winegrowers. It currently has 1300 members. "Natural", "resistant", and "heroic" winegrowers are some of the distinctions attesting to the branching out of new winemaking into many sensibilities. In a world market that is increasingly sensitive to environmental issues (Palmieri and Perito 2020) and to the cultural identities of foodstuffs, new corporate strategies include the search for old native grape varieties that survived the anti-phylloxera campaigns. So, new wine company strategies are sensitive to the rediscovery of grape varieties at risk of disappearing. Biodiversity becomes a value to be defended and passed on. Winegrowing areas participate in the "revenge of the countryside" (Barberis 2009) as a social, ethical, and ecologically balanced experience. Beyond the names, definitions and motivations, the continuous evolution of the wine cluster resulting from the introduction of differentiations and business alternatives confirms the many atypicalities of the wine sector (Patchell 2011) which, while having to adapt to an extremely fickle demand, proved capable of using the challenges to offer a wide variety of products and emotions to wine consumers.

References

Amatori, Franco, and Andrea Colli, eds. 1999. *Impresa e industria in Italia dall'Unità a oggi*. Venezia: Marsilio.

Anderson, Kym, Signe Nelgen, and Vicente Pinilla, eds. 2017. *Global Wine Markets, 1860 to 2016: A Statistical Compendium*. Adelaide: University of Adelaide Press.

Annuario statistico dell'agricoltura italiana. 1936–1938.
Atti del comitato dell'inchiesta industriale. Deposizioni scritte. Categoria 1–4 vini. 1874. Firenze: Stamperia Reale.
Baily, Samuel L. 2004. *Immigrants in the Lands of Promise: Italians in Buenos Aires and New York City, 1870–1914*. Ithaca, NY: Cornell University Press.
Banti, Alberto M. 1996. *Storia della borghesia italiana. L'età liberale*. Rome: Donzelli.
Barberis, Corrado. 2009. *Ruritalia. La rivincita delle campagne*. Rome: Donzelli.
Barrio, Patricia. 2008. El empresariado vitivinícola de la provincia de Mendoza (Argentina) a principios del siglo XX. *Historia agraria* 45: 81–111.
Bell, George. 1993. The South Australian Wine Industry 1858–1876. *Journal of Wine Research* 4: 147–163.
Berta, Pierstefano, and Giusi Mainardi, eds. 1997. *Storia regionale della vite e del vino in Italia. Piemonte*. Milan: Unione Italiana Vini.
Besana, Claudio, Rita d'Errico, and Renato Ghezzi. 2017. *Cheese Manufacturing in Twentieth Century. The Italian Experience in an International Context*. Bruxelles: Peter Lang.
Betri, Maria Luisa. 1988. L'alimentazione popolare nell'Italia dell'Ottocento. In *Storia d'Italia. Annali 13. L'alimentazione*, ed. Alberto Capatti, Alberto De Bernardi, and Angelo Varni, 7–22. Turin: Einaudi.
Biagioli, Giuliana. 2000. *Il modello del proprietario imprenditore nella Toscana dell'Ottocento: Bettino Ricasoli. Il patrimonio, le fattorie*. Firenze: Olschki.
Boletti, Pio. 1890. *Trattato popolare di enologia applicata alle cantine sociali ed al commercio del vino*. Milan: Trevisini.
Bollettino di notizie agrarie 11 (1889): 286.
Botteon, Armando, et al. 1968. *Le cantine sociali in Italia*. Rome: Luigi Scialpi editore.
Breckenridge, Saylor R., and Ian Taplin. 2005. Entrepreneurship, Industrial Policy and Clusters: The Growth of the North Carolina Wine Industry. *Research in the Sociology of Work* 15: 209–230.
Caglioti, Daniela L. 2000. Industria, organizzazione e cultura imprenditoriale nel mezzogiorno dell'Ottocento. Mélanges de l'École Française de Rome. *Italie et Méditerranée* 112: 125–150.
Canciullo, Giovanna M. 2013. Il Mezzogiorno e la Grande depressione. Capitale straniero, imprenditori e trafficanti tra crisi e trasformazione. *Studi Storici* 54: 681–699.
Cantino, Valter, Massimo Pollifroni, Damiano Cortese, and Mariangela Longo. 2016. Branding Strategies and Accounting in Wine Production: A Historical Analysis of Fontanafredda (1858–1932). In *Accounting and Food. Some Italian Experiences*, ed. Massimo Sargiacomo, Luciano D'Amico, and Roberto Di Pietra, 23–41. New York: Routledge.

Carioscia, Michele, Carlo Alberto Muratore, Nicola Rubino, Fabio Tondo, and Gianfranco Volpe. 2012. Il caso di Martini&Rossi. In *Marketing strategico e branding. Case studies*, ed. Anna Claudia Pellicelli, 273–279. Torino: G. Giappichelli.

Carlucci, Michele. 1889. Concorso a premi per vini da pasto stabilito con R. decreto 10 ottobre 1885. *Annali di agricoltura*.

Casalini, Mario. 1953. *Le cantine sociali cooperative*. Rome: Centro sociali cooperative.

Censimento degli opifici e delle imprese industriali al 10 giugno 1911. Vol. IV. Dati analitici concernenti il numero, il personale e la forza. 1914. Rome: tip. Bertero.

Charters, Steve. 2006. *Wine and Society. The Social and Cultural Context of a Drink*. London: Routledge.

Chiapparino, Francesco. 1998. "Tra polverizzazione e concentrazione. L'industria alimentare dall'Unità al periodo tra le due guerre." In *Storia d'Italia. XIII. L'alimentazione*, edited by Alberto Capatti, Alberto De Bernardi and Angelo Varni, 207–268. Torino: Einaudi.

Cinotto, Simone. 2012. *Soft Soil, Blanck Grapes: The Birth of Italian Winemaking in California*. New York: New York University Press.

———. 2013. *The Italian American Table. Food, Family and Community in New York City*. Urbana: University of Illinois Press.

Cirio, Patrizia. 1990. *Carlo e Camillo Gancia: strategie industriali 1850–1935*. Cavallermaggiore: Gribaudo.

Ciuffoletti, Zefiro. 2009. *Alla ricerca del «vino perfetto». Il Chianti del barone di Brolio. Ricasoli e il Risorgimento vitivinicolo italiano*. Firenze: Leo S. Olschki.

Coletti, Francesco. 1901. *Le associazioni agrarie in Italia dalla metà del secolo decimottavo alla fine del decimonono*. Rome: Tip. Dell'unione Cooperativa Editrice.

Colli, Andrea. 2002. *I volti di Proteo. Storia della piccola impresa in Italia nel Novecento*. Torino: Bollati Boringhieri.

———. 2012. Contextualizing Performances of Family Firms: The Perspective of Business History. *Family Business Review* 25: 243–257.

Colli, Andrea, and Mats Larsson. 2014. Family Business and Business History: An Example of Comparative Research. *Business History* 56: 37–53.

Corsi, Alessandro, Simonetta Mazzarino, and Eugenio Pomarici. 2019. The Italian Wine Industry. In *The Palgrave Handbook of Wine Industry Economics*, ed. Adeline Alonso Ugaglia, Jean-Marie Cardebat, and Alessandro Corsi, 47–76. London: Palgrave Macmillan.

Cova, Alberto. 1988. Problemi tecnici ed economici della produzione di vino in Italia tra Otto e Novecento. In *Il vino nell'economia e nella società italiana medioevale e moderna*, ed. Accademia dei Georgofili, 319–337. Firenze: Accademia economico-agraria dei Georgofili.

Dandolo, Francesco. 2010. *Vigneti fragili. Espansione e crisi della viticoltura nel Mezzogiorno in età liberale*. Napoli: Guida.

De Leo, Carolina, and Giovanni Favero. 2018. *Ca' Foscari e Carpenè Malvolti. Il Risorgimento dell'economia nel Veneto dell'Ottocento*. Venezia: Edizioni Ca' Foscari—Digital Publishing.

Dell'Osa, Dario. 2019. Viticulture and Winemaking in Abruzzo from the Unification of Italy to the Development of the Cooperation System. In *A History of Wine in Europe, 19th to 20th Centuries, Vol. I, Winegrowing and Regional Features*, ed. Silvia A. Conca Messina, Stephane Le Bras, Paolo Tedeschi, and Manuel Vaquero Piñeiro, 253–280. Cham: Palgrave Macmillan.

Enotria. Rivista Vinicola d'Italia. 1927. 10: 277.

Etudes monographiques sur la cooperation agricole dans qualques pays. vol II. Argentine, Autriche, Hongrie, Italie, Suisse. 1914. Rome: Institut International d'Agriculture.

Fauri, Francesca. 2015. *Storia economica delle migrazioni italiane*. Bologna: Il Mulino.

Federico, Giovanni. 2004. The Growth of World Agricultural Production, 1800–1938. *Research in Economic History* 22: 125–181.

———. 2009. Agriculture and Economic Development in Italy, 1870–1939. In *Agriculture and Economic Development in Europe Since 1870*, ed. Pedro Lains and Vicente Pinilla, 234–254. London: Routledge.

———. 2011. Italian Agriculture 1861–1940: A Regional Perspective. In *Growth and Stagnation in European Historical Agriculture*, ed. Patrick Svensson, Anne-Lise Head-König, and Mats Olsson, 285–307. Turnhout: Brepols Publishers.

Federico, Giovanni, and Pablo Martinelli. 2018. Italy to 1938. In *Wine Globalization. A New Comparative History*, ed. Kym Anderson and Vicente Pinilla, 130–152. Cambridge: Cambridge University Press.

Felisini, Daniela. 2019. Far from the Passive Property. An Entrepreneurial Landowner in the Nineteenth Century Papal State. *Business History* (Published Online: 22 April 2019). https://doi.org/10.1080/00076791.2019.1597853.

Fenoaltea, Stefano. 1982. The Growth of the Utilities Industry in Italy, 1861–1913. *The Journal of Economic History* 42: 601–627.

———. 2011. *The Reinterpretation of Italian Economic History: From Unification to the Great War*. Cambridge: Cambridge University Press.

Frascani, Paolo. 2012. *Le crisi economiche in Italia. Dall'Ottocento a oggi*. Bari-Rome: Laterza.

Garcia-Parpet, Marie-France. 2009. *Le marché de l'excellence. Les grands crus à l'épreuve de la mondialisation*. Paris: Seuil.

Giacalone Monaco, Tommaso. 1958. *Diagnosi della depressione economica del vino: imprenditori innovatori, tentiamo di salvare il Marsala*. Padova: Cedam.

Giannetti, Renato, and Michelangelo Vasta. 2006. *Evolution of Italian Enterprises in the 20th Century*. Heidelberg and New York: Physica-Verlag.

Gilinsky, Armand, Jr., Sandra K. Newton, and Rosana FuentesVega. 2016. Sustainability in the Global Wine Industry: Concepts and Cases. *Agriculture and Agricultural Science Procedia* 8: 37–49.

Guy, Kolleen M. 1997. Wine, Work and Wealth: Class Relations and Modernization in the Champagne Wine Industry, 1870–1814. *Business and Economic History* 26: 298–303.

Hancock, David. 2009. *Oceans of Wine: Madeira and the Emergence of American Trade and Taste*. New Haven, CT: Yale University Press.

Harsanyi, Doina P. 2005. A Resilient Elite. Survival and Decadence in the Case of the French Nobility. *Scandinavian Journal of History* 30: 286–297.

Heintz, William F. 1995. *Wine Country: A History of Napa Valley: The Early Years, 1838–1920*. Santa Barbara: Capra.

Hertner, Peter. 1984. *Il capitale tedesco in Italia dall'Unità alla prima guerra mondiale: banche miste e sviluppo economico italiano*. Bologna: Il Mulino.

Indagine sul settore vinicolo. 2020. Milano: Mediobanca.

Kovatz, Simone. 2013. The Geography of Quality Wine in United Italy Areas and Producers. In *Typicality in History. Tradition, Innovation and Terroir*, ed. Giovanni Ceccarelli, Alberto Grandi, and Stefano Magagnoli, 305–321. Brussels: Peter Lang.

L'enotecnico. Periodico di viticoltura e di enologia, 1895, n. 12: 133–135.

La cooperation agricole, vol. 1. 1931. Rome: Institut international d'agriculture.

La cooperazione nell'industria enologica, *Giornale vinicolo italiano e scientifico*, 1896, n. 26–28: 301–304.

La cooperazione vinicola italiana. 1974. Bergamo: editrice San Marco.

Lapsley, James T. 1996. *Bottled Poetry. Napa Winemaking from Prohibition to the Modern Era*. Berkeley, Los Angeles and London: University of California Press.

Law, Marc, and Gary D. Libecap. 2006. The Determinants of Progressive Era Reform. The Pure Food and Drugs Act of 1906. In *Corruption and Reform: Lessons from America's Economic History*, ed. Edward L. Glaeser and Claudia Goldin, 319–342. Chicago: University of Chicago Press.

Le Guénanff, Guylaine. 2020. Parcours de vie de vignerons toscans et prise en charge discursive. In *La vigne et le vin. Transformation des filières et des acteurs*, ed. Théodore Georgopoulos et al., 259–273. Paris: Harmattan.

Lissone, Sebastiano. 1895. *Le cantine sociali*. L'enotecnico. Periodico di viticoltura e di enologia 9: 97.

Loubère, Leo A. 1978. *The Red and the White. The History of Wine in France and Italy in the Nineteenth Century*. Albany: State University of New York.

———. 1990. *The Wine Revolution in France: The Twentieth Century*. Princeton, NJ: Princeton University Press.

Macry, Paolo. 1988. *Ottocento. Famiglia, élites e patrimoni a Napoli*. Torino: Einaudi.

Maffi, Luciano. 2010. *Storia di un territorio rurale. Vigne e vini nell'Oltrepò Pavese*. Milan: FrancoAngeli.

———. 2019. The Development of Wine Growing and Oenology in Southern Piedmont and Oltrepò Pavese. In *A History of Wine in Europe (19th–20th Centuries). Vol. I. Winegrowing and Regional Features*, ed. Silvia A. Conca Messina, Stephan Le Bras, Paolo Tedeschi, and M. Vaquero Piñeiro. London: Palgrave Macmillan.

Maffi, Luciano, Manuel Vaquero Piñeiro, and Paolo Tedeschi. 2019. Sviluppo tecnologico e nascita dell'industria del vino in Italia alla fine dell'Ottocento. In *Il settore agro-alimentare nella storia dell'economia europea*, eds. Ezio Ritrovato and Giovanni Gregorini, 195–212. Milan: FrancoAngeli.

Maffi, Luciano, Paolo Tedeschi, and Manuel Vaquero Piñeiro. 2020. Notes sur les caves sociales en Italie de la fin du XIXe siècle au fascisme. In *L'utopie au jour le jour. Une histoire des expériences coopératives (XIXe–XXe siècle)*, ed. Alexia Blin, Stéphane Gacon, François Jarrige, and Xavier Vigna, 245–255. Nancy: Arbre Bleu.

———. 2021a. A New Wine for the International Market: Italian Public Institutions' Initiatives Supporting the Oenological Sector (1870–1910). In *The World of Wine, Networks and Scales: Intermediation in the Production, Distribution and Consumption of Wine*, ed. Stéphanie Lachaud-Martin, Corinne Marache, Julie McIntyre, and Mikaël Pierre, 115–129. Bruxelles: Peter Lang.

———. 2021b. Winemakers and Wine Industry Developments in Italy (1884–1914). *Journal of Wine Research* 32 (4): 229–246.

Magagnoli, Stefano. 2015. La globalizzazione del gusto. Esposizioni universali e prodotti alimentari. *Quaderni storici* 1–2: 41–49.

Marchesi, Francesco. 1924. *Le cantine sociali.* Casale Monferrato: Marescalchi.

Mignemi, Niccolò. 2017. *Coopératives et mondes agricoles. France et Italie, 1880–1950.* Rennes: PUR.

Mora, Pierre. 2006. Key Factors of Success in Today's Wine Sector. *International Journal do Wine Marketing* 18 (2): 139–149.

———. 2016. *Wine Positioning: A Handbook with 30 Case Studies of Wine Brands and Wine Brands and Wine Regions in the World.* Cham: Springer.

Moser, Peter, and Tony Varley. 2013. The State and Agricultural Modernisation in the Nineteenth and Twentieth Centuries. In *Integration through Subordination: The Politics of Agricultural Modernisation in Industrial Europe*, ed. Peter Moser and Tony Varley, 13–40. Turnhout: Brepols.

Maffi, Luciano, Vaquero Piñeiro Manuel, and Paolo Tedeschi. 2019. Sviluppo tecnologico e nascita dell'industria del vino in Italia alla fine dell'Ottocento. In *Il settore agro-alimentare nella storia dell'economia europea*, eds. Ezio Ritrovato, Giovanni Gregorini, 195–212. Milano, FrancoAngeli..

North, Douglas C. 1990. *Institutions, Institutional Change and Economic Performance.* Cambridge: Cambridge University Press.

Nützenadel, Alexander. 2008. A Green International? Food Markets and Transnational Politics, c. 1850–1914. In *Food and Globalization. Consumption,*

Markets and Politics in the Modern World, ed. Alexander Nützenadel and Frank Trentmann, 153–171. New York: Berg.

Nuvolari, Alessandro, and Michelangelo Vasta. 2017. The Geography of Innovation in Italy, 1861–1913: Evidence from Patent Data. *European Review of Economic History* 21 (3): 326–356.

Nuvolari, Alessandro, Pier Angelo Toninelli, and Michelangelo Vasta. 2018. What makes a Successful (and Famous) Entrepreneur? Historical Evidence from Italy (XIX–XX Centuries). *Industrial and Corporate Change* 27 (3): 425–427.

O'Keefe, Kerin. 2014. *Barolo and Barbaresco. The King and Queen of Italian Wine.* Oakland, CA: University California Press.

Oddy, Derek J., and Alain Drouard, eds. 2016. *The Food Industries of Europe in the Nineteenth and Twentieth Centuries.* London and New York: Routledge.

Official Catalogue of Exhibitors. Universal Exposition St. Louis. 1904. St. Louis: Committee on Press and Publicity by the Official Catalogue Company.

Palmieri, Nadia, and Maria Angela Perito. 2020. Consumers Willingness to Consume Sustainable and Local Wine in Italy. *Italian Journal of Food Science* 32 (1). https://itjfs.com/index.php/ijfs/article/view/1648.

Pan-Montojo, Juan. 1994. *La bodega del mundo. La vid y el vino en España (1800–1936).* Madrid: Alianza.

———. 2001. Las industrias vinícolas españolas: Desarrollo y diversificación productiva entre el siglo XVIII y 1960. In *Viñas, bodegas y mercados: El cambio técnico en la vitivinicultura española, 1850–1936*, ed. James Simpson, Juan Carmona, Josep Colomé, and Juan Pan-Montojo, 313–334. Zaragoza: Prensas Universitarias de Zaragoza.

Patchell, Jerry. 2011. *The Territorial Organization of Variety. Cooperation and Competion in Bordeaux, Napa and Chianti Classico.* London and New York: Routledge.

Paul, Harry W. 1996. *Science, Vine and Wine in Modern France*, 1996. Cambridge: CUP.

Pedrocco, Giorgio. 1994. Un caso e un modello: viticoltura e industria enologica. In *Studi sull'agricoltura italiana. Società rurale e modernizzazione*, ed. Pier Paolo D'Attorre and Alberto De Bernardi, 315–343. Milan: Feltrinelli.

Petrick, Gabriella M. 2012. Industrial Food. In *The Oxford Handbook of Food History*, ed. Jeffrey M. Pilcher, 258–278. Oxford: Oxford University Press.

Pilbeam, Pamela M. 1990. *The Middle Classes in Europe 1789–1914. France, Germany, Italy and Russia.* London: Macmillan.

Pilcher, Jeffrey M. 2017. *Food in World History.* New York: Routledge.

Pinilla, Vicente, and Maria-Isabel Ayuda. 2002. The Political Economy of the Wine Trade: Spanish Exports and the International Market. *European Review of Economic History* 6: 51–85.

Pinilla, Vicente, and María-Isabel Ayuda. 2007. The International Wine Market, 1850–1938: An Opportunity for Export Growth in Southern Europe. In *Wine*

Society and Globalization. Multidisciplinary Perspectives on the Wine Industry, ed. Gwyn Richard Campbell and Nathalie Guibert, 179–200. New York: Palgrave Macmillan.

Planas, Jordi. 2016a. State Intervention in Wine Markets in the Early 20th Century: Why was it So Different in France and Spain? *Revista de Historia Economica / Journal of Iberian and Latin American Economic History* 35: 175–206.

———. 2016b. The Emergence of Winemaking Cooperative in Catalonia. *Business History* 58: 264–282.

Quinn, Martin, and João Oliveira, eds. 2019. *Accounting for Alcohol. An Accounting History of Brewing, Distilling and Viniculture*. New York: Routledge.

Ramón Resina, Joan, and William Viestenz. 2012. *The New Ruralism. An Epistemology of Transformed Space*. Madrid: Iberoamericana Veruvert.

Regan-Lefebrvre, Jennifer. 2017. John Bull's Other Vineyard: Selling Australian Wine in Nineteenth-Century Britain. *The Journal of Imperial and Commonwealth History* 45: 259–283.

Santini, Cristina, Alessio Cavicchi, and Leonaro Casini. 2013. Sustainability in the Wine Industry: Key Questions and Research Trends. *Agricultural and Food Economics* 1: 9. On-line: https://agrifoodecon.springeropen.com/articles/10.1186/2193-7532-1-9; https://doi.org/10.1186/2193-7532-1-9.

Schumpeter, Joseph. 1934. *The Theory of Economic Development*. Cambridge, MA: Harvard University Press.

Segers, Yves, Jan Bieleman, and Erik Buyst, eds. 2009. *Exploring the Food Chain: Food Production and Food Processing in Western Europe, 1850–1990*. Turnhout: Brepols.

Simpson, James. 2000. Cooperation and Cooperatives in Southern European Wine Production. *Advances in Agricultural Economic History* 1: 95–126.

———. 2005. Cooperation and Conflicts: Institutional Innovation in France's Wine Markets, 1870–1911. *Business History Review* 79: 527–558.

———. 2011. *The Emergence of a World Industry, 1840–1914*. Princeton: Princeton University Press.

Suffia, Ilaria, Andrea M. Locatelli, and Claudio Besana. 2018. Cheese trademarks: Italian dairy firms' practices during the 20th century. *Business History* 60 (8): 1227–1254.

Swinnen, Johan F.M. 2010. Agricultural Protection Growth in Europe 1870–1969. In *The Political Economy of Agricultural Price Distortions*, ed. Kim Anderson, 141–162. Cambridge: Cambridge University Press.

———. 2018. *The Political Economy of Agricultural and Food Policies*. New York: Palgrave Macmillan.

Tedeschi, Paolo. 2003. Il rinnovamento colturale. Aspetti della viticoltura bresciana fra Ottocento e Novecento. In *La civiltà del vino: fonti, temi e produzioni vitivinicole dal medioevo al Novecento*, ed. Gabriele Archetti. Brescia: Centro culturale artistico di Franciacorta e del Sebino.

———. 2019. The Noble Entrepreneurs Coming from the Bourgeoisie: Counts Bettoni Cazzago during the Nineteenth Century. *Business History* (Published Online: 16 September 2019). https://doi.org/10.1080/00076791.2019.1653283.

Teughels, Nelleke, and Peter Scholliers, eds. 2017. *A Taste of Progress: Food at International and World Exhibitions in the Nineteenth and Twentieth Centuries*. London and New York: Routledge.

Toniolo, Gianni. 1990. *An Economic History of Liberal Italy*. London: Routledge.

———. 2013. An Overview of Italy's Economic Growth. In *The Oxford Handbook of the Italian Economy since Unification*, ed. Gianni Toniolo, 13–16. Oxford and New York: Oxford University Press.

Unioncamere. 2007. Rapporto sul settore vitivinicolo. Milan: Unioncamere.

Unwin, Tim. 1996. *Wine and the Vine: An Historical Geography of Viticulture and the Wine Trade*. London and New York: Routledge.

Vaquero Piñeiro, Manuel. 2018a. Il vino della Campania tra XIX e XX secolo: aziende e mercati internazionali. *Archivio Storico per le Province Napoletane* 136: 191–204.

———. 2018b. La nascita delle cantine industriali in Italia. In *Paesaggi della protoindustria. Luoghi e processi della produzione dalla storia al recupero*, ed. Augusto Ciuffetti and Roberto Parisi, 205–217. Rome: Carocci.

———. 2019. Italian Wine in Argentina (1861–1914): Market, State Intervention and Wineries. *Storia Economica* 22: 331–360.

Vaquero Piñeiro, Manuel, and Luciano Maffi. 2018. The Diffusion of Italian Wine in the United States (1861–1914). La difusión del vino italiano en los Estados Unidos (1861–1914). *Rivar. Revista Iberoamericana de viticultura, agroindustria e ruralidad* 15: 176–196.

Vasta, Michelangelo. 1999a. Capitale umano e ricerca scientifica e tecnologica. In *Storia d'Italia. Annali 15. L'industria*, ed. Franco Amatori, Duccio Bigazzi, Renato Giannetti, and Luciano Segreto, 1041–1124. Torino: Einaudi.

———. 1999b. *Innovazione tecnologica e capitale umano in Italia (1880–1914). Le traiettorie tecnologiche della seconda rivoluzione industriale*. Bologna: il Mulino.

Vrontis, Demtris, Stefano Bresciani, and Elisa Giacosa. 2016. Tradition and Innovation in Italian Wine Family Businesses. *British Food Journal* 118 (8): 1883–1897.

Wolikow, Serge. 2019. History of a Vineyard in Champagne: From Eighteenth to Twenty-First Century. In *A History of Wine in Europe (19th–20th Centuries). Vol. I. Winegrowing and Regional Features*, ed. Silvia A. Conca Messina, Stephan Le Bras, Paolo Tedeschi, and Manuel Vaquero Piñeiro, 51–76. Cham: Palgrave Macmillan.

CHAPTER 4

Production, Trade, and Consumption

The difficult process of structural renewal of the Italian wine sector began in the 1880s: although there was a great resistance to change, the expansion of international trade and the transformation of wine into one of the main products of the globalisation of consumption pushed the Italian wine producers to become more competitive in outside Italy. During the following decades, Italy did not lose its status as a major wine producer, but it was not enough to have large quantities of grapes available to become a relevant winemaker country. On the contrary, the abundant harvests contributed to the backwardness of the sector: many Italian winemakers did not improve their products because they had the possibility to sell low-quality wine in the internal market, and they had easy sales for their good wines destined to cut foreign ones. In this regard, it is interesting to highlight that historiography is analysing these phenomena in other states, such as France and the Iberian Peninsula (Becuwe et al. 2022; Ferrer-Alos 2021). For the Italian wine sector, there existed a strong correlation between production trends, external trade, and the evolution of internal demand: only exogenous factors such as diseases concerning vines as well as relevant changes in the international markets could modify the winemakers' attitudes.

Between 1887 and 1890 two crises marked the Italian wine sector: the customs war with France (James and O'Rourke 2013) and the outbreak of the Argentine financial crisis. Both ruptures strongly reduced Italian exports of blending wines and de facto obliged many winemakers to

M. Vaquero Piñeiro et al., *A History of Italian Wine*,
https://doi.org/10.1007/978-3-031-06097-7_4

abandon their old oenological practices and to modernise them to obtain better quality and more competitive wines. This progressive process of modernisation, which concerned a few share of wine producers, was interrupted by the two World Wars and the policy of Fascism Regime concerning the use of grapes. Wars strongly reduced the purchasing power of the Italian people, diminishing the market for better quality but more expensive wines; fascism regime gave more importance to the cereals and preferred table grapes. Only in the 1960s, thanks to the economic miracle, the new European policies, the recourse to new technologies, and due to the great increase of the Italian wine quality with good quality/price ratio, it allowed winemakers to bring the image and the sales of their products in the international markets. The quality improvement continued until the early new millennium, and Italian wine exports became one of the material (and also immaterial) pillars of the success of the "made in Italy" in the world (Fortis and Sartori 2016). Finally, in the last two decades, Italy has been at the top of the world rankings, concerning the quantity and quality of vines and wines: the great gap existing with France in the nineteenth century and until the 1960s was definitively filled.

4.1 Production

Considering the available sources on the trend of wine production in Italy in the first decades after the birth of the Kingdom of Italy (1861), it is possible to collect the following aspects (Ottavi 1893): first official data on Italian wine production were in fact indicated by the Ministry of Agriculture in the late 1870s (Federico 2003; D'Onofrio 2016; Tedeschi and Vaquero Piñeiro 2018). At the end of the nineteenth century, the surface cultivated with vines was about 3.4 million hectares, three times more than in Spain and France where vineyards respectively occupied 1.6–1.8 million hectares. However, the difference concerning the wine production was inferior, even if in a context where data were not always reliable. While the first official data for 1878–1879 indicate an annual production for Italy of around 18.7 million hectolitres, other sources in fact put the production capacity at 26–30 million hectolitres: this last appears a much more reliable estimate and considered the effects of diseases concerning vines. The destruction caused in France by phylloxera created for some years the conditions for the massive increase of Italian grapes exports (Pinilla and Ayuda 2007; Anderson, Nelgen, and Pinilla 2017): the phylloxera infection started in Italy in 1879 with only a few hectares (24.58), and the affected area

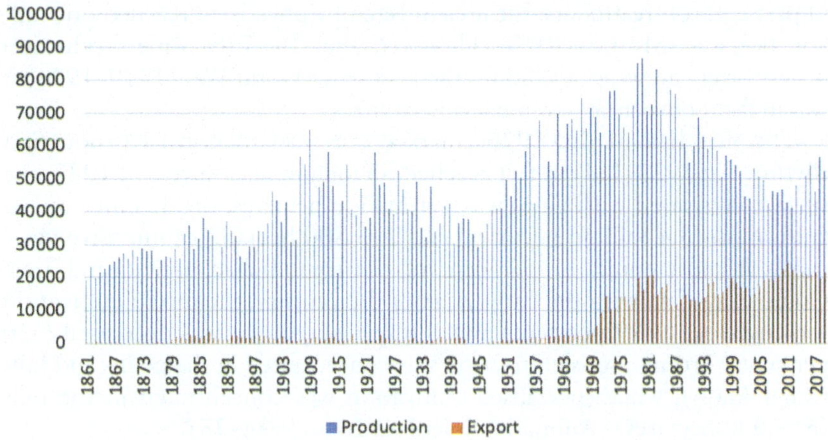

Fig. 4.1 Production and export of Italian wine, 1861–2019 (thousands of hectolitres). (Source: Serie storiche Istat)

increased from 1884, involving in particular the Sicily (Dandolo 2010). When at the end of the nineteenth century, the municipalities involved by the phylloxera became 816 for a surface of 350,948 hectares, the evolution of ampelographic knowledge allowed to realise an efficient contrast by the action of the anti-phylloxera consortia grafting local vines with the American ones. So, Italian wine production could reach 65 million hectolitres in 1909 (Fig. 4.1). Also the quality improved: the average level remained low, but the increase in the production costs to face the vines diseases obliged winemakers to justify the prices which necessarily increased. Besides, many vines having lowest yields were eliminated, in particular where they could be substituted with cereals and meadows (as in the plain): the remaining vines normally allowed to have better quality grape harvests.

The First World War upset the balance of the European production fabric, including the wine sector (Simpson 2011; Unwin 1993; De Ferrière Le Vayer and Williot 2013). Italian wine production during the war years was around 39 million hectolitres, while in 1913–1914 it fluctuated between 47 and 58 million hectolitres. The impact was very negative, in 1915 (the first year of the conflict), the wine production plummeted to 19 million hectolitres. During the war, most productive regions (Apulia, Sicily), although far from the battlefields, suffered the loss of male labour and the abandonment of pest control treatments: the area infected by phylloxera reached one million hectares (Mortara 1925). However, after 1915 the wine production progressively grew up: including the post–First World War (1919–1924) it was in fact between 39–48 million hectolitres.

The first half of the 1920s marked a return to normal production rhythms: in 1923, Italy in fact reached 53 million hectolitres. In 1925, the surface cultivated with vines was 4.4 million hectares, divided into mixed cultivation (3.4 million hectares) and 745,800 hectares of intensive vineyards: it represented 17 per cent of the total Italian agricultural area. Comparing data with the early years of the century, the mixed cultivation was increased (+29 per cent) and the specialised one was decreased (-20 per cent). Promiscuous cultivation was concentrated in the plains and hills (2.8 million), while specialised cultivation was concentrated in the hills (568,000 hectares) (Annuario vinicolo d'Italia 1924–25).

New changes arrived during the economic crisis of the 1930s: there was a relatively moderate downturn, demonstrating that wine grape production in Italy did not depend on market trends only. In 1929, specialised vines occupied 935,000 hectares, and mixed ones, 2,900,000 hectares for a total production of 74 million quintals of grapes. The production of table grapes was 170,000 tons (Fig. 4.2). In the first half of the 1930s, promiscuous cultivation maintained the previous surface, while the specialised one significantly decreased (742,000 ha): total grape production

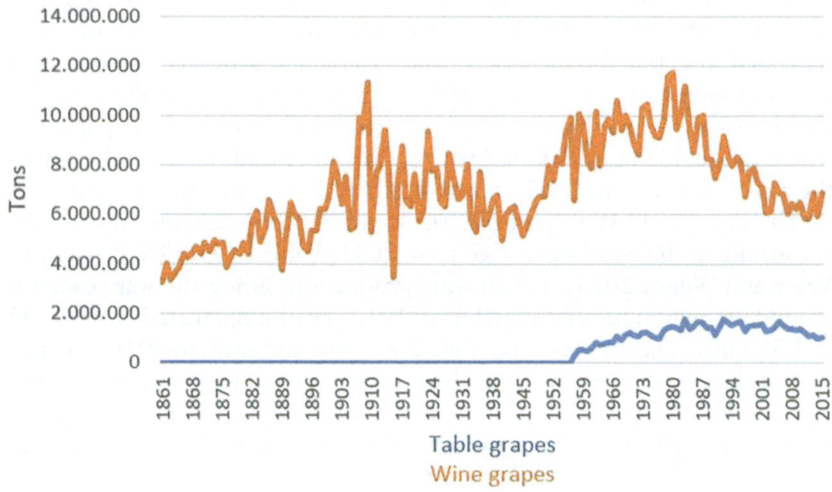

Fig. 4.2 Production of table and wine grapes (tons). (Source: Serie storiche Istat)

was 6.3 million tons and 38.9 million hl of wine; table grape exports reached 30,000 tons and wine exports, 920,000 thousand hl. In the following years, vines in promiscuous cultivation remain at 2,900,000 ha, while specialised cultivation grew up around 900,000 ha. Total grape production was 5–7 million tons, and production of table grapes arrived at 100,000 tons; total wine production arrived at 38,500,000 hl; export of table grapes 243,000 quintals and wine exports, 1,500,000 hl. Grape production per hectare was 350–370 kilograms.

The main consequences of this division were: the production of large quantities of low-quality wine; a deep-rooted tendency towards conservatism among the small producers, for whom what mattered was to ensure a plentiful harvest to the detriment of any change or improvement. This caused a low level of Italian wine exports compared to the huge amount of wine produced and consumed locally. A survey carried out in 1942 showed that there were wine vats in the country with a capacity of about 45 million hectolitres, about 82 per cent belonged to small and medium-sized wineries, and only 3–4 per cent of the vats were located in great and modern ones. The remaining 15 per cent belonged to "enopoli", that is cooperatives where local grapes and wines were collected and sold (Casalini 1953).

The most novel element of the 1930s was the intense legislative action carried out by the Fascist regime: Italian wine production suffered the negative world economic situation and, in addition, the effects of the policies imposed by Fascism. During the 1920s, it paid greater attention to cereal-growing and heavily penalised other agricultural sectors, and it decided to revaluate the Italian currency ("Quota 90") making Italian products less competitive in the international markets. So, during the 1930s, the world negative economic trend and the related reduction of prices on wine market decreasing revenues prevented wine producers from tackling upgrading and modernisation of their plants. Between 1914 and 1939, per capita wine consumption in Italy fell from 124 to 89 litres per year. The autarchy de facto increased the bureaucratic apparatus concerning all agricultural sectors without improving them (Fabiani 2015, 153). For the viticulture, the decreasing revenues and the related reduction of investment in new technologies resulted in the return of the phylloxera problem: in 1932, there were many municipalities with phylloxera or suspected infection (1145 in Lombardy, 925 in Piedmont, 187 in Tuscany, and 118 in Marche).

In the 1940s, the prices strongly reduced (Fabiani 1987, 98): they were at 12 lire per quintal during the Second World War, and they collapsed to 4 lire in 1950. Most winemakers continued to be anchored to a purely quantitative logic, and the production abundance reduced prices in markets in a context where wine consumption continued to decrease. They preferred to produce large quantities of poor-quality wine, and there was no national policy favouring the changes needed to realise more competitive products: the increasing amount of unsold wine and the cellars full pushed winemakers to sell at a very low price to make alcohol, and this was a great disincentive to invest in quality improvement. Besides, balancing production to adapt it to real demand was a difficult solution to implement because it clashed with most winemakers' traditional attitude: to increase their production from year to year (Brassley et al. 2016). Also the agricultural contracts were not conducive to the improvement of the productive methods: where sharecropping agreements prevailed, farmers wanted to produce more and more following a logic of accumulation that was difficult to eradicate and far from the existing consumers' demand.

In this context, the renewal of the Italian viticulture was helped by international aid plans and the European integration process obliging the Italian economy to enter in a new more competitive Western European market (Fauri 2010; Fauri and Tedeschi 2011). In 1949, the need to sign a customs convention with France introduced the perspective of policies improving the quality of wines for export. The Italian government was now in favour of signing the Madrid Agreement of 1891 on the international registration of trademarks. This strongly modified a sector accustomed to operating outside international rules, thus encouraging the proliferation of false names, fraud, and counterfeiting. In the new Western Europe made by the cold war, it was necessary to introduce improvements in the Italian wine sector and to abandon the attitude in favour of large and low-quality production.

However, the absence of adequate national policies and the predominance of a mass of small producers hindered the improvement of the wine sector (Oddy and Drouard 2016; Segres et al. 2009). In 1949, there were only 15,000 medium-sized wineries: they represented 40 per cent of national production, and the remaining part depended on small wineries, more than 300,000. These type of wineries kept traditional an expensive productive methods and due to this aspects, they were not able to modernise their production. The only positive elements were represented by the progressive mechanisation of the cellars and the creation of an auxiliary industry specialising in the manufacture of winemaking machinery: they were mostly exported and this put in evidence the backwardness and the lack of investments in the Italian wine sector.

Only during the 1950s, some changes were slowly introduced: the new high production costs caused by rising wages, electricity prices, transport costs, and also the tax burden obliged winemakers to change; the production of a litre of wine went from a cost of 5–6 Italian lire to 7–8 lire (+37 per cent), and this was not compatible with a wine market where these were the sale prices. The quantitative logic did not allow most small winemakers to survive, and new trends imposed by both the international markets and the technological innovations concerning cellars and winemaking modified wine producers' attitudes and favoured the increase of the specialised vines surface (Brassley 1997; Swinnen 2018; Stranieri and Tedeschi 2019).

Private wine cooperatives, in the mid-twentieth century, were an expedient by which the state intervened in the wine sector (Planas 2016; Simpson 2000; Fernández 2014). Faced with the weakness or even absence of producer associations such as those that had been founded in France, cooperatives, formed by small farmers, not only helped to solve some of the many problems that were fossilising Italian agriculture, but also acted as a transmission belt between political power and the interests that revolved around wine (Medina-Albaladejo and Menzani 2017).

Exports to Europe continued to increase, favoured by the communities of Italian workers settled in Germany, France, and Belgium. By the end of the 1950s, exports to France increased by 53.3 per cent, from 22 million dollars to 35 million dollars (1959). This demonstrates the close relationship between migration flows and food exports. The increase in exports to European countries also explains the growth in the area dedicated to growing table grapes, which reached 90,000 hectares, almost all of which were concentrated in the southern regions, especially Apulia and Sicily. Exports of fresh table grapes peaked in 1983 at 1.7 MT.

Besides, in the mid-1950s, the Italian wine sector faced a serious crisis due to a weather event. In February 1956, after a terrible frost that lasted several weeks, more than 50 per cent of the vines in some regions were destroyed, and the national production fell from 62 to 42 million hectolitres (-32 per cent). It was a disaster for the fragile economy of sharecroppers and peasants' families because their income significantly depended on the sale of wine (and olive oil). However, this event obliged the government to pay more attention to the wine sector: using the *Ispettorati*

Provinciali di Agricoltura, it introduced a new massive vine renewal programme replacing the damaged vines with new ones (Garrabou Segura 2005). In many terroirs, this replacemnt improved the development of grape varieties and encouraged the end of promiscuous cultivation with a general improvement in grape quality and an increase in yields. In addition, viticulturists received non-repayable public loans and more attention was dedicated for the graduation of new oenologies having the task to educate winemakers to introduce new efficient productive techniques as well as to progressively develop a new entrepreneurial mindset in Italian terroir.

The introduction of much younger vines into the Italian countryside provoked a new overall production in the following years. Only a part of winemakers in fact decided to produce less wine improving its quality, the others continued to follow the traditional attitude.

The introduction of new grape varieties selected with a view to achieving an increase in productivity provoked the growth in production up to 1980 when the 86 million hectolitres were exceeded, even with a grape yield per hectare of over 10,000 kg of grapes (Fig. 4.3). Between 1950 and 1980, the grape yield per hectare in Italy increased by almost 130 per cent. Wine production in Italy increased by 33 per cent from the 1950s to 1980s, that is from 54 to 72 million hectolitres (Corsi et al. 2018). Only in the 1990s, the wine production began to decline and decreased to 47 million hectolitres between 2010 and 2016.

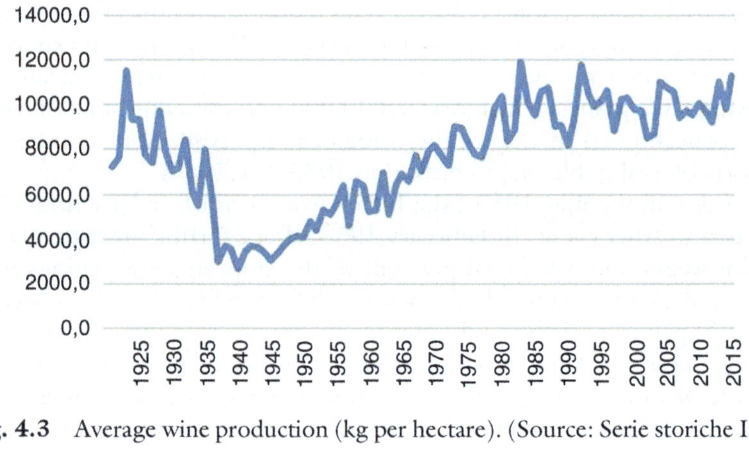

Fig. 4.3 Average wine production (kg per hectare). (Source: Serie storiche Istat)

Table 4.1 Surface of mixed and specialised (1909–1970)

Years	*Mixed vines*		*Specialised vines*	
	Surface (millions of hectares)	*Grape production (millions of tons)*	*Surface (millions of hectares)*	*Grape production (millions of tons)*
1909–10	3.4		0.919	
1929	2.9		0.935	
1932–35	2.9		0.742	
1936–42	2.9		0.965	
1946–48	2.9		0.981	
1949–50	2.8		1	
1951–54	2.8	2.8	1	5.1
1956–59	2.7	2.9	1.1	6.3
1968–70	1.6	2.4	1.3	8.7

Source: ISTAT

The promiscuous cultivation of vines, which remained stationary at around 3 million hectares until the end of the 1950s, reduced to 1.3 million hectares in the 1960s. The specialised viticulture moved very slowly around the million hectares. Specialised vine cultivation predominated in Apulia and Sicily until the 1950s, when the massive entry of wine production from Veneto and Emilia-Romagna took place (Table 4.1). In 1968, these four regions accounted for 47.7 per cent of the wine grape production in specialised cultivation in Italy. In opposite, Piedmont and Tuscany remained outside the areas with an increase in specialised viticulture. The new vines and wine productions did not show an improvement in the quality: there was the production of large masses of blended wine to be sent outside, in particular in Apulia and Sicily where this continued until the end of the twentieth century (Corsi et al. 2018).

The division between promiscuous and specialised cultivation in fact hid a complex reality (Federico and Martinelli 2018). In 1911, specialised vineyards occupied an area of 852,000 hectares, a quarter of the vines under promiscuous cultivation which reached 3.5 million hectares. Considering the wine production, the specialised vineyards contributed for 21 million hectolitres and the promiscuous one for 25 million ha (Serpieri 1930): the productivity of the specialised vineyards more than compensated the difference in surface. This continued until 1969, the last year in which the traditional division between "specialised" and "promiscuous" cultivation was used in the statistical data: the promiscuous one covered 1.6 million hectares and produced 2.4 million tons of grapes; the specialised one covered 1.1 million hectares and produced 8.8 million tons of grapes (79.5 quintals per hectare). In 1969, the total Italian grape production for wine reached 11,634 million tons: one million tons of table grapes were also produced.

In 1970, vineyard register introduced a new classification: "main cultivation" and "secondary cultivation". The former accounted for 1.2 million hectares and the latter for 704,999 hectares. The application of this criterion resulted in the area classified in the land register as "secondary cultivation" being considerably smaller than the area previously recorded as "mixed cultivation". This radical change had main reasons: for example, many areas that previously were already classified as mixed or promiscuous land were reported as "area with scattered plants". The new cadastre denounced the previous overestimation of the surface of vineyards: prior to 1970, official data on vine cultivation had a high margin of discretion, as shown by the difference between the vineyard extension recorded by the new cadastre (1,895,510 hectares) and the previous one (2,773,100 hectares). In 1970, when the first Italian vineyard register was drawn up, there were 1.9 million farms in Italy having wine grape vines; there were only 78,388 farms having table grape vines. There was a total predominance of holdings with less than one hectare of vines: they were 1.3 million. In terms of regions, Veneto, Sicily, and Apulia were at the top with 9.8, 9.7, and 9.3 per cent respectively of the national surface area dedicated to the production of wine and table grapes (Fig. 4.4).

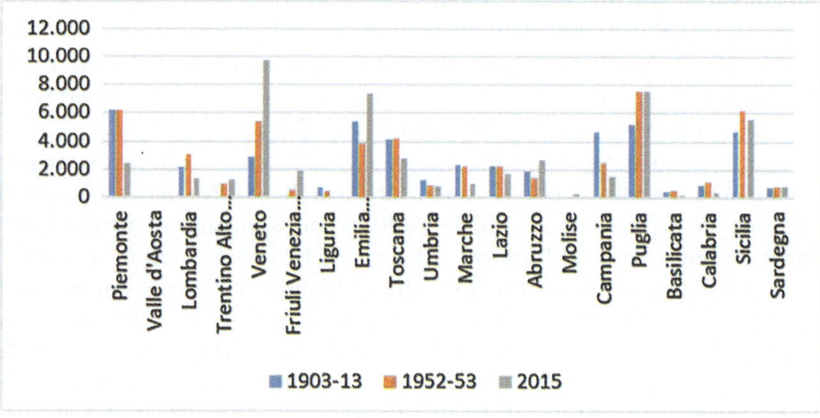

Fig. 4.4 Wine production divided by regions (2015). (Source: Ismea)

In the 1980s, "main crop" vines remained at 1.2 million and "secondary crop" declined to 470,659 hectares. Total grape production exceeds 11.7 million tons with yields of 92 quintals per hectare. The transition of Italian viticulture towards a full modernisation was not easy: the wine production continued to be very high provoking low prices and excess pro duction. European attempts to find a balance between supply and demand in Italy were met with a cold shoulder by winemakers: the same happened for new rules against fraud and adulteration (Bonardi 2014). The export trend to non-European markets like the United States or the Soviet Union increased, but it was more and more wider the difference between wineries selling on the market normal table wines and other one selling higher quality wines: as better winemakers tried to gain market share, the reaction of other ones was a further reduction of prices which obliged to cut the productive costs or to make frauds damaging customers and all the wine sector.

This explains why during the 1980s, the Italian wine sector highlighted its "excellence" and "worst". In 1983, the DOC Bolgheri was created and established the Sassicaia as one of the most important Italian wine, but which became the symbol of the new high oenological line of Italian wines. With the "Superitalians" or "Supertuscans" defined as "one of the most refreshing and innovative" products which "more than any others, contributed to raising the quality of Italian wines", the creation of a luxury wine entrepreneurship became a reality (Resnick 2020). In 1986, the tragic methanol wine scandal broke out: some criminals decided to add too much methanol in their wine to increase the alcohol content and reduce the production costs. This provoked blindness and neurological damages for consumers and killed 23 people: the sales of Italian wine immediately collapsed and in the following months all Italian winemakers were forced to search for new quality and safety standards.

In the 1980s, relevant changes in European Common Market pushed viticulturists to change their attitudes: the entry into ECM of three countries producing much wine (Greece in 1981, Portugal and Spain in 1986) in fact imposed the need for a new regulation aimed at reducing the European countries' production. European institutions granted premiums for the permanent abandonment of land and funding for the restructuring and renewal of vineyards. Between 1988 and 1996, 137,788 hectares of vineyards were abandoned in Italy.

In 1990, the wine plan was approved in Italy and production stood at 53 million hectolitres, of which 8.7 million (16.3 per cent) were VQPRD wines: however, the following years were marked by strong instability related to the elimination of protections on the ECM and production had a significant drop. The GATT agreements of 1994, which guaranteed the arrival in Europe of large quantities of competitive wines from Argentina, Australia, South Africa, and other non-European countries: besides, the negotiations with Bulgaria, Hungary and Romania, countries with a solid wine sector, allowed their products a privileged access to the ECM. Finally, new European regulations prohibited the blending of European wines with other important ones: this obviously created problems for winemakers who produced wines for blending. European regulations also limited surpluses through uprooting: winemakers instead called for financing aid to be used to convert vineyards considered too old (as it happened in regions such as Tuscany).

In the new millennium, there was a substantial stability, but their distribution by geographical area changed: the decline of the main European producers (they fell from 48.7 per cent in 1990 to 43.4 in 2005) was contrasted by the growth of the surface of vines in countries representing main competitors (Australia, New Zealand, Chile, and South Africa) as well as in China and the United States. The fall in the surface for vines was very marked in European countries having a strong winegrowing tradition: France lost 6.2 per cent, while in Portugal, Italy, and Spain the reduction was around 20 per cent. Italian vines surface from 1,024,282 hectares in 1990 to 837,845 in 2005. The world wine production increased by 9.7 per cent from 1995 to 2005: at the same time, the decline of traditional European producers was relevant and Italy passed from 5.6 million tons in 1995 to 4.9 in 2005 (-11 per cent). The reduction in the wine production of France (which remained the leading world producer) and Italy (passed from 22.2 per cent of the world production in 1995 to 17.6 per cent in 2005) was offset by Spain, which doubled its share coming very close to the production capacities of the two European leaders (UnionCamere 2007).

The increase in wine consumption in the new millennium depended on the growth and consolidation of non-European markets as well as the United Kingdom one (it grew by 30.1 per cent from 2000 to 2005, and it arrived at 11.9 million hectolitres). The wine consumption in the main European markets fell (-5.5 per cent in France and -10.4 per cent in Italy) because of the gradual shift away from daily consumption towards one of a more occasional nature, even if more attentive to the quality.

Turning to production volumes, the production of wine grapes in Italy, in 2005, amounted to approximately 71 million quintals, a decrease compared to 2004, but an increase compared to previous years. Production was concentrated mainly in four regions (Puglia, Sicily, Veneto, and Emilia-Romagna). Apulia was the first region for production (11.8 million quintals), followed by Sicily (9.8 million), Veneto (9.7 million), and Emilia-Romagna (8.7 million). Italian production of wine and must in liquid state was in 2005 at 50.6 million hectolitres, with an average annual increase of 4.3 per cent compared to 2002. Most of the production was obviously concentrated in the four regions previously identified for the more relevant grape production. The most significant increases in 2002 were recorded in Apulia (+14.3 per cent per year), Piedmont (+9.5 per cent), and Valle d'Aosta (+9.4 per cent). On the contrary, the production decreased in Lazio (-6.2 per cent), Basilicata (-4.8 per cent), Liguria (-3.4 per cent), and Abruzzo (-3.1 per cent).

The Italian wine sector confirmed its very fragmented production structure: this continued to prevent the achievement of economically efficient dimensions and the exploitation of economies of scale, but it did not avoid further improvement of the quality of the production. However, the small size of the production fabric had a negative impact on promotional policies: in the early years of the new millennium, there were 108,808 farms producing DOC-DOCG grapes with an average size of just 2.15 hectares. The production of DOC and DOCG wine was 13.9 million hectolitres. The regions with the greatest number of winemakers were Apulia and Sicily, with more than 30,000 active wineries, followed by Piedmont, Emilia-Romagna, Veneto, and Abruzzo, with between 13,000 and 15,000 ones. The small size of farms also concerned companies dedicated to the processing and packaging of wine. In 2005, there were 1994 active enterprises with an average of only 9 employees per establishment. The weaknesses resulting from the small size of the companies were compensated by a high varietal richness and a high added value of the products.

The reduction of vines surface continued in the second half of the 2000s (-7 per cent): it was the effect of the increase of vineyards for the production of DOP wines (+40 per cent) and the decrease of vineyards for other wines (-31 per cent). The winemakers' choice to improve the quality of their productions was clearly showed by data from the statistical surveys.

At the end of the first decade of the twenty-first century, the area devoted to the production of wine and table grapes averaged 600,000 ha (Table 4.2). Obviously, there was a strong predominance of wine grapes

Table 4.2 Vines surface in hectares (2000–2010)

	2000	*2010*	*Variation %*
Total vines	717,334	664,296	- 7
Grapes for wine and table	715,521	663,005	- 7
Grapes for wine	675,545	625,7	- 7
DOP wines	233,605	320,859	+ 40
Other wines	441,94	304,841	- 31
Table grapes	39,976	37,305	- 8

Source: ISTAT

over table grapes, while the division between grapes for PDO wines and "other wines" was essentially equal. In terms of regions, Puglia and Sicily dominate with 107,000 and 114,000 hectares respectively, followed by Veneto (78,000 hectares), Emilia-Romagna (60,000 hectares), Tuscany (60,000 hectares), Piedmont (47,000 hectares), and Lombardy (23,000 hectares). These regions accounted for 73 per cent of the area dedicated to vine cultivation. Comparing these data with the situation existing in the 1960s, when promiscuous viticulture (2.7 million hectares) predominated over specialised one (one million), allows to note that mixed vine-growing almost disappeared while specialised vine-growing occupied even less land than it did in the 1930s. At the same time, a significant increase in yields per hectare cultivated was registered: from 52 quintals in 1950 to 112 quintals in 2015.

The Italian production of PDO and PGI wine in 2016 was almost 25 million hectolitres, of which 22.5 million were bottled, corresponding to 3 billion bottles (+5.4 per cent compared to 2015). The evolution of the production of quality wines marked a further step forward compared to the previous year. In 2016, 14.5 million hectolitres were certified as PDO (DOC and DOCG). The production of PDO wines was driven by Prosecco, which, with over 3 million hectolitres, increased by 15.5 per cent compared to 2015. At the same time, PGI wine reached 10.4 million hectolitres, of which one was exported in bulk. The production of quality wines, in spite of the large number of awards spread throughout the country, remained very concentrated. The top three regions, in order Veneto, Piedmont, and Tuscany, produced 57 per cent of the total certified PDO. There was also a strong concentration of PDO: the top ten accounted for 53.5 per cent of total production by volume and 57 per cent by value. In the case of PGI, the concentration was even greater: the top

Table 4.3 Main DOP wines by value of their certified product (millions of euros)

DOP wines	*2014*	*2015*	*2016*	*2017*
Prosecco	267	467	629	631
Conegliano Valdobbiadene	120	140	161	184
Chianti	110	102	87	81
Asti	122	100	103	107
Chianti classico	53	100	112	98
Amarone	98	98	83	103
Alto Adige	84	94	82	100
Barolo	65	73	79	89
Valpolicella Ripasso	64	68	63	86
Others	896	964	1043	1105
Total	1916	2243	2510	2655

Source: ISTAT

10 represented more than 80 per cent of the total in terms of both volume and value.

In 2018, the last year for which statistics are available before Covid-19 changed the general picture, world wine production was 260 million hectolitres and Italy was the leading producer with a share of 18.6 per cent of the world total. The value of Italian production was around 16 billion euros. PDO wines (DOC and DOCG) accounted for 42.2 per cent of the total, while PGI wines for 24.9 per cent and ordinary wines for the remaining 32.9 per cent (Mediobanca 2020). Table 4.3 shows the weight of the different Italian PDO wines. Prosecco accounted for 17 per cent of exports in terms of volume and 20 per cent of value in millions of euros (Pomarici et al. 2019). Immediately after there were some classic wines such as Chianti, Asti, Barolo, Soave, and Valpolicella together with others of more recent consolidation such as Montepulciano d'Abruzzo, Alto Adige, and Trentino (Table 4.3). So, new wines were together with a small group of historic DOC wines (Chianti, Barolo, Asti) that emerged at the end of the nineteenth century and then, going through relevant transformations, maintained a high level of prestige on international markets.

So, the Italian wine sector changed the area dedicated to vines and its wine production system. These changes were related to various factors, starting with technological and scientific innovations, the socio-economic and political conditions of the country, and ending with cultural aspects, especially relating to per capita wine consumption and, in more recent years, awareness of sustainable development and the inclusion of vineyards in an overall discussion of landscape and environmental quality. The growth of organic wine moved in this direction. In Italy, as in other countries, organic wine slowly consolidated its share (Jones and Grandjean 2017) and did not enjoy a good image among traditional producers and consumers who considered it a product of lesser prestige. The area cultivated by organic viticulture grew from 52.4 to 105,000 hectares between 2010 and 2017 (more than 100 per cent). It concentrated the production in Apulia and Sicily, creating an interesting identification between organic wine and Southern Italy.

4.2 International Trade

The wine trade, especially foreign trade, always represented a fundamental element for the wine sector (Roccas 2003): Italian wine exports did not improve until the end of the 1870s (Federico 1991). In 1862–1863, wine accounted for just 2.2 per cent of agricultural exports, far behind raw silk and even olive oil (Federico 1982, 2003). From 1871, when 227,000 hectolitres were shipped abroad, bulk wine exports increased to one million hectolitres in 1879 and then 3.5 million in 1887. These were the years in which the growth of sales to France, hit by phylloxera, introduced a new element into the Italian wine sector: as a consequence of the leap forward in wine exports, in percentage terms (5.9 per cent) the wine in fact reached the third position among primary products sent abroad, a position it maintained until the end of the nineteenth century (Federico et al. 2011, 25–26; Becuwe et al. 2022) (Fig. 4.5).

Together with the growth of foreign sales of bulk wine in barrels, there was also an increase in exports of bottled wine. Official data did not distinguish table wine from other exported beverages such as vermouth, marsala and bitters, but the French market surely helped the increase in exports of barrelled wines (Pedrocco 1994). French wine imports rose from just one million to seven million hectolitres and reached 12 million hectolitres at the end of the 1880s. Because of the massive French imports of blended wines, the total value of Italian wine reached 1300 million lire, ahead of wheat and olive oil. For the years 1886–1890 the average value of wine exports exceeded 65.5 million lire, a long way from the silk trade (273 million) but at the same level as olive oil (66 million) (*Notizie e studi sull'agricoltura* 1892, LXIV). In terms of the volume of wine produced and the surface of vines, Apulia led the regions in the supply of blending wine for France. In the provinces of Bari and Lecce, blended wines

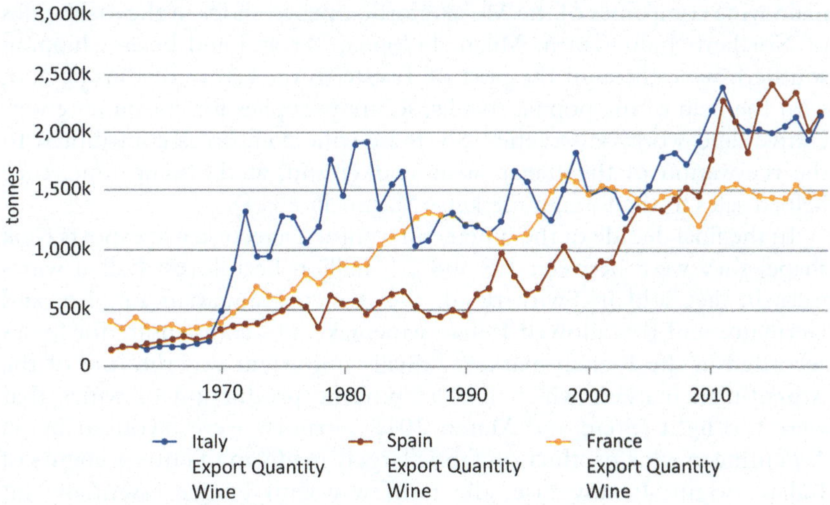

Fig. 4.5 Italy, Spain, and France export quantity wine (1961–2019). (Source: FAOSTAT)

accounted for 59–62 per cent of total wine production. Important French wineries planted thousands of hectares of common vines in Apulia, even imposing the type of vines best suited to French wine production (Ritrovato 2010).

The decline occurred between 1889 and 1891 because of the Franco-Italian tariff war, which had an extremely negative effect on the wines of Southern Italy. Italian wines were in fact quickly replaced on French market by the Spanish, Portuguese and Algerian wines (Meloni and Swinnen 2018), but the sudden closure of the French market for Italian blending wines was a heavy blow, mainly for the colonists and small producers who had made cultivation choices in the boom years attracted by the easy and secure earnings (Chevet et al. 2018). In any case in Southern Italy, after the tightening of customs tariffs in 1887, the production for the foreign market did not disappear: between 1893 and 1894, wine in barrels totalling 2.5 million hectolitres was exported to the ports of Southern Italy (Bari, Brindisi, Naples, Catania, Palermo, Trapani) (Bulletin of the General Company 1895). At the beginning of the twentieth century, Apulia accounted for 11 per cent of total Italian wine production, equivalent to 5 million hectolitres. With the end of the season of easy exports to France, main wine companies of the Mezzogiorno opened shops in the main cities of Northern Italy (Turin, Milan, Bologna, Verona) and began shipping southern wine through the port of Trieste to the centre of Europe and, with the help of the popular banks, secured supplies for consumer cooperatives and workers' societies. So, these wine companies contributed to the renovation of the Italian wine market and, at the same time, they helped sales to the foreign markets (Martinelli 2014).

In the final decade of the nineteenth century, a recovery in exports took shape: they were between 1.5 and 2.5 million hectolitres: Italian wines were in fact sold in Switzerland, the Austro-Hungarian Empire, and Germany, and this allowed Italian winemakers to compensate for losses recorded in the French market. Equally important was the role of the Argentinian market, which had a domestic production of wines that were too light (Stein and Mateu 2018): exports were favoured by an Argentinean wine production fabric largely made up of entrepreneurs of Italian origin. In any case, the tariff war with France essentially hit exports of bulk wine in barrels, and had a minimal effect on the positive trend enjoyed by bottled wine, which rose from 17,000 hectolitres in 1885 to 33,700 in 1893 (Notizie e studi intorno ai vini ed alle uve d'Italia (Roma, 1896): 844–845): it was a small fraction of total exports as bottled wine was destined for a clientele more willing to pay higher prices for quality imported wines. Data were underestimated: the customs statistics office began to distinguish separately the qualities of exported wines only in 1896 and, moreover, it was diffused the practice of sending wine abroad in bulk in barrels to be bottled locally, which reduced transport costs and paid less customs duty. This practice in particular concerned exports in American markets and in the United Kingdom: bulk wine was transformed into bottled wine "with an elegant and attractive appearance that made them more acceptable to the consumer". This completely escaped official data avoiding to calculate the actual dimension in hl.

In the early twentieth century, exports fell below 1.5 million hectolitres and further declined to reach their lowest point in 1906: 800,000 hl (Pedrocco 1994). Nevertheless, there was a substantial recovery in the following years due to a sharp contraction in prices: in 1914 exports exceeded 2 million hl. With the First World War, Italian wine exports decline again and only in 1919, due to French losses, they came back to the previous levels.

With regard to the value of those exports, consistent data are only available after 1905 when exports reached 36.7 million Italian lire. This value remained stable until 1910 when the export of 2 million hectolitres allowed to realise commercial revenues of 86 million lire, opening a period of relative stability that lasted until 1914. In 1915, Italy faced the collapse of 53 per cent in terms of sales: however, the increase in prices allowed exports to reach 140 and 285 million lire respectively for the years 1917–1918, a value also depending on the great inflation rate caused by the war.

Looking at the US market, in 1861 Italian wine exports amounted to scarcely 4 hectolitres: however, during the following decades the situation improved, with exports averaging around 2400 hectolitres, and peaks of 5135 (1866) and 5790 (1872). The first relevant change in the trend came at the end of the 1870s: exports reached 33,016 hectolitres and increased during the 1880s and early 1890s when, between 1885 and 1893, were about 63,000 hl. Even if during the last years of the nineteenth century a drastic fall arrived (18,282 hl in 1897), exports progressively increased in the twentieth century until the First World War: they amounted to 65,812 hectolitres in 1905 and, moreover, reached 100–150,000 hectolitres during the period 1910–1914. The exports of bottled wines rose from barely 5000 in 1874 to 250,000 in 1881 and 629,000 in 1890. The main Italian wines exported to the United States were Barbera, Chianti, Barolo, Capri, Lacryma Christi, Asti sparkling wine, and Moscato di Siracusa, as well as the traditional Marsala and Vermouth (Hancock 2009; Macedo 2011).

Even in a scenario characterised by continuous growth, Italian remained less competitive: European wine imports in New York, in the years 1889–1891, were in fact dominated by Germany, Hungary, and France, while in the US market also Spanish wines were better positioned than Italian ones. Besides, Italian wines had to face the new Californian wine which met the demand for low-quality wines. The US wine production amounting to about 4728 hectolitres in 1840 reached 920,000 hectolitres during the 1870s, and 70 per cent of Californian wine was sold in New York, Chicago, Saint Louis, and New Orleans. So, Italian wines had to compete both with European wines (in particular French and German brands) and US ones. The presence of a huge population of Italian immigrants represented a favourable factor for the sales of Italian wines (Rosano 2000), but it was not enough, and Italian winemakers needed to improve the quality of their products as well as commercial strategies.

Besides, at the beginning of the twentieth century, American custom policies became particularly strict with regard to alcoholic beverages: in addition to suffering the negative consequences of the prohibitionist propaganda limiting the consumption of alcohol (Hames 2012), there was the promulgation of the "Pure Food and Drugs Act" in 1906, which dealt with substances harmful to the public health and the marketing of food products (Rumbarger 1989) (Image 4.1).

The content in sulphur dioxide had to be indicated on the label of the bottle, even if the quantities were lower than the limits set by the regulations (Rossati 1910). This legislation, aimed at ensuring food quality, discouraged the purchase of wines with such an uninviting label. The federal authorities' decision in particular hit Chianti (Mocarelli 2013): it had some success among the American élite, but its alcohol content had to be lowered to 12 per cent. It also had to adapt to the taste of the country, as American customers were looking for smooth, less austere, and palatable wines (Ottavi and Marescalchi 1897). New limits in fact obliged best Italian winemakers, who were open to innovation and followed the advice of the Oenotechnic station, to adapt their product according to the health and food hygiene legislation, reducing the rate of alcohol and ensuring the quality of the bottled product: new US rules de facto improved the quality of Italian wines.

If bottled wine exports indicated the ability of Italian winemakers to carve out a space in the quality wine segment, exports of bulk wine in barrels involved a more complex scenario: the bulk wine in fact met the demand of Italian migrants in the United States (Malpezzi and Clements 1999) and also represented the perfect tool to avoid tariffs. High duties on Italian bottled sparkling wines were in fact imposed by the trade agreement signed on 8 February 1900: a duty of 9.60 dollars for a dozen bottles was an exceptionally burdensome measure, which effectively closed the US market to sparkling Muscat Canelli. The response was to step up bulk exports of Muscat in barrels, to be bottled and labelled in the United States. One of the plants specialising in the bottling of Italian Muscat was the Pleasant Valley Wine Company, in Rheims, New York (Pellechia 2015; Pinney 1989). In this way, Italian winemakers got around the customs restrictions and continued to take advantage of the expansion of the US market. The absence of stringent regulations on designations of origin and food authenticity allowed Italian producers of Muscat Canelli to bottle and sell their products in the United States.

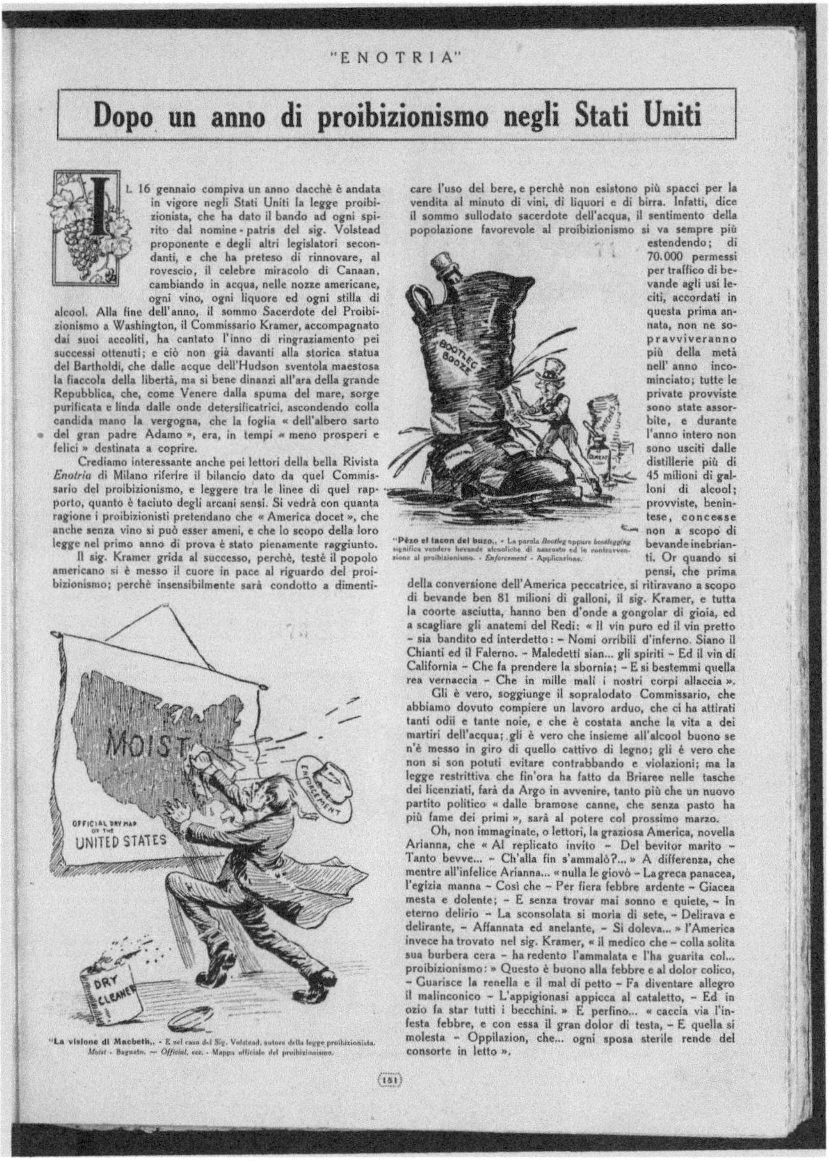

"ENOTRIA"

Dopo un anno di proibizionismo negli Stati Uniti

IL 16 gennaio compiva un anno dacchè è andata in vigore negli Stati Uniti la legge proibizionista, che ha dato il bando ad ogni spirito dal nomine-patris del sig. Volstead proponente e degli altri legislatori secondanti, e che ha preteso di rinnovare, al rovescio, il celebre miracolo di Canaan, cambiando in acqua, nelle nozze americane, ogni vino, ogni liquore ed ogni stilla di alcool. Alla fine dell'anno, il sommo Sacerdote del Proibizionismo a Washington, il Commissario Kramer, accompagnato dai suoi accoliti, ha cantato l'inno di ringraziamento pei successi ottenuti; e ciò non già davanti alla storica statua del Bartholdi, che dalle acque dell'Hudson sventola maestosa la fiaccola della libertà, ma si bene dinanzi all'ara della grande Repubblica, che, come Venere dalla spuma del mare, sorge purificata e linda dalle onde detersificatrici, ascondendo colla candida mano la vergogna, che la foglia « dell'albero sarto del gran padre Adamo », era, in tempi « meno prosperi e felici » destinata a coprire.

Crediamo interessante anche pei lettori della bella Rivista *Enotria* di Milano riferire il bilancio dato da quel Commissario del proibizionismo, e leggere tra le linee di quel rapporto, quanto è taciuto degli arcani sensi. Si vedrà con quanta ragione i proibizionisti pretendano che « America docet », che anche senza vino si può esser ameni, e che lo scopo della loro legge nel primo anno di prova è stato pienamente raggiunto.

Il sig. Kramer grida al successo, perchè, testè il popolo americano si è messo il cuore in pace al riguardo del proibizionismo; perchè insensibilmente sarà condotto a dimenticare l'uso del bere, e perchè non esistono più spacci per la vendita al minuto di vini, di liquori e di birra. Infatti, dice il sommo sullodato sacerdote dell'acqua, il sentimento della popolazione favorevole al proibizionismo si va sempre più estendendo; di 70.000 permessi per traffico di bevande agli usi leciti, accordati in questa prima annata, non ne sopravviveranno più della metà nell' anno incominciato; tutte le private provviste sono state assorbite, e durante l'anno intero non sono usciti dalle distillerie più di 45 milioni di galloni di alcool; provviste, benintese, concesse non a scopo di bevande inebrianti. Or quando si pensi, che prima della conversione dell'America peccatrice, si ritiravano a scopo di bevande ben 81 milioni di galloni, il sig. Kramer, e tutta la coorte asciutta, hanno ben d'onde a gongolar di gioia, ed a scagliare gli anatemi del Redi: « Il vin puro ed il vin pretto – sia bandito ed interdetto: – Nomi orribili d'inferno. Siano il Chianti ed il Falerno. – Maledetti sian... gli spiriti – Ed il vin di California – Che fa prendere la sbornia; – E si bestemmi quella rea vernaccia – Che in mille mali i nostri corpi allaccia ».

"Pèzo el tacon del buzo,, - La parola *Bootleg* oppure *bootlegging* significa vendere bevande alcooliche di nascosto ed in contravvenzione al proibizionismo. - *Enforcement* - Applicazione.

"La visione di Macbeth,, - E nel caso del Sig. Volstead, autore della legge proibizionista. *Moist* - Bagnato. — *Official, ecc.* - Mappa ufficiale del proibizionismo.

Gli è vero, soggiunge il sopralodato Commissario, che abbiamo dovuto compiere un lavoro arduo, che ci ha attirati tanti odii e tante noie, e che è costata anche la vita a dei martiri dell'acqua; gli è vero che insieme all'alcool buono se n'è messo in giro di quello cattivo di legno; gli è vero che non si son potuti evitare contrabbando e violazioni; ma la legge restrittiva che fin'ora ha fatto da Briaree nelle tasche dei licenziati, farà da Argo in avvenire, tanto più che un nuovo partito politico « dalle bramose canne, che senza pasto ha più fame dei primi », sarà al potere col prossimo marzo.

Oh, non immaginate, o lettori, la graziosa America, novella Arianna, che « Al replicato invito – Del bevitor marito – Tanto bevve... – Ch'alla fin s'ammalò?... » A differenza, che mentre all'infelice Arianna... « nulla le giovò – La greca panacea, l'egizia manna – Così che – Per fiera febbre ardente – Giacea mesta e dolente; – E senza trovar mai sonno e quiete, – In eterno delirio – La sconsolata si moria di sete, – Delirava e delirante, – Affannata ed anelante, – Si doleva... » l'America invece ha trovato nel sig. Kramer, « il medico che – colla solita sua burbera cera – ha redento l'ammalata e l'ha guarita col... proibizionismo: » Questo è buono alla febbre e al dolor colico, – Guarisce la renella e il mal di petto – Fa diventare allegro il malinconico – L'appigionasi appicca al cataletto, – Ed in ozio fa star tutti i becchini. » E perfino... « caccia via l'infesta febbre, e con essa il gran dolor di testa, – E quella si molesta – Oppilazion, che... ogni sposa sterile rende del consorte in letto ».

151

Image 4.1 Prohibition in the United States and Italian wines. (Source: Rossati (1921), n. 5, p. 151)

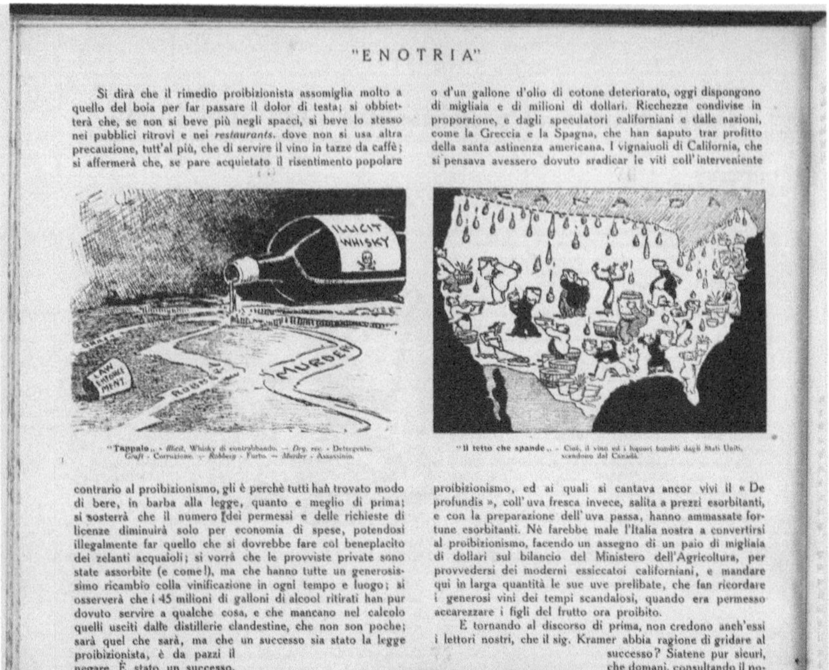

"ENOTRIA"

Si dirà che il rimedio proibizionista assomiglia molto a quello del boia per far passare il dolor di testa; si obbietterà che, se non si beve più negli spacci, si beve lo stesso nei pubblici ritrovi e nei *restaurants*, dove non si usa altra precauzione, tutt'al più, che di servire il vino in tazze da caffè; si affermerà che, se pare acquietato il risentimento popolare contrario al proibizionismo, gli è perchè tutti han trovato modo di bere, in barba alla legge, quanto e meglio di prima; si sosterrà che il numero dei permessi e delle richieste di licenze diminuirà solo per economia di spese, potendosi illegalmente far quello che si dovrebbe fare col beneplacito dei zelanti acquaioli; si vorrà che le provviste private sono state assorbite (e come!), ma che hanno tutte un generosissimo ricambio colla vinificazione in ogni tempo e luogo; si osserverà che i 45 milioni di galloni di alcool ritirati han pur dovuto servire a qualche cosa, e che mancano nel calcolo quelli usciti dalle distillerie clandestine, che non son poche; sarà quel che sarà, ma che un successo sia stato la legge proibizionista, è da pazzi il negare. È stato un successo, senza dubbio, il proibizionismo; un successo colossale, all'*American style*.

Io credo, per esempio, e mi ci giocherei un soldo italiano al cambio della giornata, che mai in vita loro nè il sig. Kramer, nè tutto il politicume democratico d'ispettori e controllori, nè i rivenditori all'ingrosso, nè i farmacisti, nè i produttori di pozioni medicinali, nè i chimici ed i barbieri, giacchè ora si bevono anche le acque profumate, nè i fabbricanti di accessorii per l'imbottigliamento dei generi proibiti, nè i carettieri ed automobilisti, nè i venditori di uva passa e di uva fresca, ecc., avessero mai sognato le ricchezze, che han conseguito in così breve tempo, in così ingente misura e con tanta poca fatica. Gente che un anno addietro non godeva la fiducia dell'anticipo d'una forma di formaggio canadese muffito, o d'un gallone d'olio di cotone deteriorato, oggi dispongono di migliaia e di milioni di dollari. Ricchezze condivise in proporzione, e dagli speculatori californiani e dalle nazioni, come la Grecia e la Spagna, che han saputo trar profitto della santa astinenza americana. I vignaiuoli di California, che si pensava avessero dovuto sradicar le viti coll'interveniente proibizionismo, ed ai quali si cantava ancor vivi il « De profundis », coll'uva fresca invece, salita a prezzi esorbitanti, e con la preparazione dell'uva passa, hanno ammassate fortune esorbitanti. Nè farebbe male l'Italia nostra a convertirsi al proibizionismo, facendo un assegno di un paio di migliaia di dollari sul bilancio del Ministero dell'Agricoltura, per provvedersi dei moderni essiccatoi californiani, e mandare qui in larga quantità le sue uve prelibate, che fan ricordare i generosi vini dei tempi scandalosi, quando era permesso accarezzare i figli del frutto ora proibito.

"Tappalo" - *Illicit*. Whisky di contrabbando. — *Dry, ecc.* - Detergente. *Graft* - Corruzione. — *Robbery* - Furto. — *Murder* - Assassinio.

"Il tetto che spande" - Cioè, il vino ed i liquori banditi dagli Stati Uniti, scendono dal Canadà.

E tornando al discorso di prima, non credono anch'essi i lettori nostri, che il sig. Kramer abbia ragione di gridare al successo? Siatene pur sicuri, che domani, consultando il povero Pantalone, il quale infine paga sempre le spese delle frenesie astute dei grossi pescicani, si volesse tornare all'antico, ad un sol uomo insorgerebbe la turba farisaica degli acquaiuoli per reclamare il rispetto delle gloriose leggi proibizionistiche, ed il Kramer, col Bacco in Toscana, minaccerebbe un'altra volta « Chi l'acqua beve - Mai non riceve - Grazie da me ». Molto più che, coi guadagni che loro il proibizionismo assicura, essi non si fanno mancare il liquido che « Così grato in seno piove - Ch'ambrosia e nettar non s'invidia a Giove ».

Oh, sia laudato il proibizionismo americano; esso che fa rifiorire il sentimento religioso tra'popoli, e che somministra, anche se condito con molto sale e moltissimo pepe, un farmaco unico ed infallibile per tutti i malanni del corpo e le infermità dello spirito.

"Il Proibizionismo in opera"

(152)

Image 4.1 (continued)

"ENOTRIA"

Quarantacinque milioni di galloni di alcool consumati in un solo anno negli Stati Uniti a scopo *religioso* o *medicinale*, vuol dire, per chi sa appena di calcolo, che non solo i sacerdoti han dovuto dir messa, ma tutti gli « abitator de la faceta terra », topi inclusi e cacàrocci, che s'è bevuto tanto di Ferro China, quanto forse non si beverebbe d'acqua in

"Le vie del contrabando"

un anno. Lo stesso popoletto, che non è alla portata di comprar carissimo il gallone di vino, ed il bicchiere refrigerante di birra, ne ha tratto il vantaggio di esser diventato chimico, nella propria cucina, al punto da sapersi fabbricare il caratello di mosto atto alla fermentazione, e la damigiana di birra. In alcune di queste istruzioni popolari, ho letto coi miei occhi, darsi alla fine il monito di non lasciar fermentare il prodotto per più di tre minuti, onde evitare una birra con più del mezzo per cento di alcool, ciò che è contrario alla legge, che va rispettata da ogni buon cittadino.

E poi, bisogna esser giusti, la violazione non infirma la legge; e si sa che sempre i pesci piccoli, che non hanno il potere di romper la rete, volentieri restano presi; ma bisogna pure far giustizia, che più umani di quello che sono, i custodi della legge, non potrebbero desiderarsi. Un pugno di miserabili dollari appiana qualsiasi difficoltà legale, e se si è condotti in *Domo Petri*, gli è proprio che si ha cattiva volontà. È stato colto, è vero, qualche povero diavolo che voleva far passare dell'alcool camuffato nelle latte di olio; qualche altro che, distratto, vendeva a gente senza permesso alcool falsificato da compare; qualche altro pure, che trasportava delle casse di *whisky*, col debito funebre accompagnamento, dentro il carro mortuario, che avrebbe dovuto servire a ricettare un cadavere; ma o gli è che è stata loro fatta la spia da amici di cui credevano di potersi fidare, o che, cocciuti, non han voluto addolcire il giusto risentimento dei custodi addetti a sorvegliare l'osservanza della legge.

Non è permesso, è vero, l'uso dei vini, come bevanda, ma per uso medicinale si vende tanto *whisky* e *cognac* quanto basta a mandare all'altro mondo un reggimento di anime. E poi, l'alto concetto americano, è più lo spirito della legge che vuole rispettato, che la legge stessa. Purchè non si venda del vino, senza pagare lo scotto, poi si commerci pure liberamente col mosto non fermentato e fermentabile, col mosto in polvere, e colla conserva d'uva in boatta, che rappresenta il *dernier cris*, più o meno *a surprise*, in fatto di vinificazione. Senza queste vedute larghe dei proibizionisti, come si farebbe a calmare i rimprocci delle coscienze di tutti noi, che di fronte al codice siamo più o meno criminali? Del resto anche a vendere quei componenti, non è mica vender del vino; o se anche venda quel vino o quella birra, se non ha un grado alcoolico rispettabile, è come l'amore senza baci, direbbe l'on. Direttore di questa degnissima Rivista.

Che se, per malignare, si voglia portare innanzi le statistiche della cresciuta criminalità, la registrazione delle sbornie più fenomenali in tempo di proibizionismo, dell'esuberante uso di narcotici in sostituzione dell'alcool, delle diminuite entrate alla cassa governativa, ciò non toglie al successo decantato del proibizionismo in azione. Le male lingue ci sono state, e ci saranno sempre; e coi nuovi rigori pel venturo anno, Mr. Kramer si propone un successo ancora più strabiliante, molto più se al corpo dei custodi dell'ordine sarà aggravata, come ne è stato fatto progetto, la Polizia nazionale, composta in gran parte d'Irlandesi e di Tedeschi, nemici dichiarati (?) del *whisky* e della birra, come lo sono gli italiani del vino e del cicchetto.

E la nuova pudica Naiade, rigenerata dall'acqua, più pura e più bella balzerà dalle spume frementi, non propriamente nella divisa troppo sfacciata dell'Innocenza, ma in... mutandine da bagno... decenti, lo sguardo avido rivolto al grappolo dell'uva bambina (*baby grape*), e circondata d'agenti protettori della vergine beltà, cui contano i passi e sorvegliano i vasi dell'infecondo talamo.

New York GUIDO ROSSATI.

"Il vino sequestrato che scorre per le vie di Los Angeles" - Fotografia

(153)

Image 4.1 (continued)

Italian wine exports also grew up in Argentina (Vaquero Piñeiro 2019). The Latin American country, destination of a massive Italian emigration, offered excellent conditions for growth to the international trade of wine coming from Italy: a large part of the population were in fact inclined to consume goods coming from the mother country. Italian wines searched for new space in a market where French table wines predominated and were easy to find in hotels and in fashionable restaurants, many of them managed by French chefs and owners (Balmaceda 2016): even if the so-called Bordeaux wine was actually the result of mixtures made with Spanish, Italian, and Southern French wines, that is it did not have all the characteristics of genuine Bordeaux wine, it in fact had a great success among a varied clientele who came from the villages of the Pampas (Correa and Wibaux 2000, 78) to the inhabitants of the cities attracted by the lifestyles existing in the European high society during the "belle époque" (Losada 2009).

So, the Argentine market contributed to changing the best Italian winemakers' attitudes: it stimulated the shift from the production of wines for local consumption or cutting to a more advanced business approach that involved learning how to take advantage of the sale of table wine in international markets. The aim to gain Argentine market shares outside traditional national circuits forced Italian winemakers to become familiar with marketing techniques and commercial advertising as well as to form effective distribution chains by involving agents and companies specialised in the transfer and sale of goods. The best Italian winemakers had to definitively separate themselves from the traditional agriculture in order to acquire the characteristics of entrepreneurs operating in a modern wine sector and having the same knowledge concerning the productive systems and the marketing that characterised entrepreneurs involved in industrial and commercial sectors: to stay in the Argentine market, it was in fact necessary to improve the quality of the wine and use new commercial strategies.

For the 1860s and 1870s, data allow to distinguish the exports of bulk wine in barrels from those in bottles: between 1861 and 1874 wine in barrels exports to Latin America accounted for a small part, as the wine fed a trade flow directed almost exclusively towards European countries (France, Austria, England, and Switzerland). On an annual average of 284,250 hectolitres of wine exported in barrels, Latin America was scarcely equivalent to 9269 hectolitres (3.2 per cent) (Vaquero Piñeiro 2019). The situation was better for bottled wines: in 1872, of a total of over 2 million bottles of Italian wine exported, the Latin American countries received 262,300 bottles, 11.75 per cent. However, it is important to note that official statistics on the export of bottled beverages did not refer solely to table wine as they also included other widely consumed alcoholic beverages such as vermouth, grappa, Marsala, and spirits.

Between 1874 and 1885, the exports of Italian wine in barrels to South America went from 8777 to 24,555 hectolitres with a clear predominance to Uruguay (Vaquero Piñeiro 2019). In 1886, exports for 51,887 hectolitres were reached and then a real leap forward took place from 1887 (121,406) to 1889 (328,779). In 1890, exports returned to 239,358 hectolitres and then to 344,492 hectolitres in 1892. Concerning Italian bottled wine exports to South America, they fluctuated in the period 1874–1887 between 4000 and 10,000 hundreds of bottles (the hundreds of bottles was the unit of measure) and experienced a vigorous growth up 14,000–18,000 hundreds of bottles between 1888 and 1890. In the early 1890s, exports reduced to 7–10,000 hundreds of bottles (Vaquero Piñeiro 2019).

Thanks to the establishment of the Stazione Enotecnica of Buenos Aires, the data concerning exports of Italian wine to Argentina were more accurate and confirmed the increase in exports since the late 1880s (Rayes 2018). Remembering that the information sometimes provided one figure for the entire range of alcoholic beverages, the level of exports of bulk Italian wines to Argentina was below 20,000 hectolitres between 1871 and 1878. Successively, the exports increased up to 60,000 hectolitres in 1887. In terms of percentage of the total Italian wine exports, from 1871 to 1886 exports to Argentina accounted for just 3–4 per cent of the total. The turning point was in 1888 and 1889 when 361,899 hectolitres were reached, respectively corresponding to 18 and 25 per cent of the total Italian wine exports. An increase coinciding with the customs trade war with France and the related loss of the main market for cutting wines and the other consequences with other powerful suppliers of cutting wine like Algeria (Meloni and Swinnen 2018). Argentina replaced France because it was in full demographic growth and consumption propelled by immigrants. So, the internationalisation of Italian agro-food products at the end of the nineteenth century was not linked to migration flows only (Zanoni 2018). At the end of the 1880s, the Argentine market was largely unknown and precise information was lacking (Fonseca 1891). Most of the winemakers sent their products over the ocean without criteria, flooding the Argentine market with several brands, names and different qualities to the detriment of a few wine producers who understood that the new market demanded specific strategies.

The years 1889–1891 were negatively marked by the Baring Crisis (Mitchener and Weidenmier 2008), a crisis caused by the failure of the historic London bank that imposed on many Latin American countries the use of drastic policies of return and expenditure restraint (Cortés Conde 1979). In 1890, there was a sharp fall in imports and an increase in prices due to the monetary turbulence generated by the collapse of the Argentine *peso* compared to the value of gold (della Paolera and Grandes 2007; Regalsky 2006; Ford 1969). In fact, between 1889 and 1890, the index of food prices went from 122 to 159 per cent and remained around 166.6 per cent until the end of the 1890s (Vence Conti and Martín 2016, 292). As a result of a climate of great uncertainty, the news sent to Italy from Argentina indicated a dramatic deterioration in living conditions (Devoto 2001; Fauri 2015) and, as a result, a reduction in consumption, including wine. In particular, the bulk wine market was damaged by the measures taken by the authorities to counter the drastic reduction in customs revenues. Before the crisis, the customs duty that was paid for Italian wine in barrels was 8 cents a litre, after it increased to 20 cents. There was also an increase in customs duties on bottled wine (from 25 to 50 cents per bottle), but it concerned common wines only: tax on high-quality wines unchanged. The immediate reaction of importing businesses was to clear much of the wine stored in port warehouses. In any case, the market response, even in the presence of a scenario dominated by currency depreciation, was quite positive for high-quality Italian wines sold at relatively high prices. The Barolo cash register remained quite stable around 12–15 *pesos nacionales*, a price that rose to 22 *pesos nacionales* for Barolo Mirafiore and to 30 *pesos nacionales* for Barolo Cinzano. The segment of bottled wines, less affected by tax increase, gave evidence of being able to adapt to the crisis.

The most serious situation occurred in the sector of imports of cut and common wines: they became easy ground for adulteration by resorting to the use of substances of dubious origin mixed with poor-quality wines. In a market subject to numerous market fluctuations, for wine in barrels there was a 57 per cent rise in *pesos nacionales* between January (75.06 p/n) and September 1890 (118.1 p/n), an upward trend that continued in 1891 when there was a progressive increase in prices. In the first semester of 1891, the "common" Italian wine ranged from 125 and 160 p/n while the common bulk French wine imported into Argentina was between 160 and 170 p/n.

In the early 1890s, the economic recession in Argentina ended up favouring selective and substitution processes concerning the wine imports (Barbero and Rocchi 2003). Until 1890, the Argentine wine market was predominated by French wines: in 1888, out of a total of 82 million litres of bulk wine, France guaranteed 41 per cent, Spain 32 per cent, and Italy 17 per cent. Concerning the bottled wine sector percentages changed but France (51 per cent) remained at the top, and it was followed by Spain (20 per cent), Germany (11 per cent), and Italy (3.4 per cent). After the crisis Spain, but especially France, significantly decreased their wine exports to Argentina: French wine in barrels went from 39 million to 29 million litres. Instead Italian winemakers were able to adapt their products to the new economic trend. When the crisis broke out in August 1890, the French wine sector was grappling with the damage caused by phylloxera and its wines were too expensive for the Argentine rich families too. These latter considered the most accredited brands of the Piedmontese wines (as Cinzano), an excellent lower price alternative. Between 1895 and 1905, the total imports of wine in barrels and bottles in Argentine (Barbera, Barolo, Ischia, Capri, etc.) each year averaged 5,789,818 gold pesos (Chiaromonte 1906, 5): Italian wine in barrels averaged 216,687 hectolitres for a value of 2,211,256 gold *pesos* and with 90,806 dozens of bottles of wine for a value of gold *pesos* of 436,775. These values corresponded, respectively, to 45.83 per cent of the imported wine in Argentina and 45.79 per cent of the monetary value of common and bottled wines. The success of Italian wine exports to Argentina was accompanied by the good results obtained in Brazil, which could not be explained solely by taking into account the Italian immigrants' demand. Italian winemakers in fact shifted production to table wines for direct consumption, much more profitable, abandoning cutting ones. They improved the quality of their products also adapting them to consumers' tastes. At the end of the nineteenth century, Italian exports of common wine in barrels to Argentina reached 2,418,089 gold *pesos*: it was the second Italian product after cotton fabrics amounting to 3,179,011 gold *pesos*. At the same time, Italian bottled wine reached 100,000 gold *pesos* (Vaquero Piñeiro 2019).

In the first years of the twentieth century, exports of Italian wine to Argentina ranged from a minimum of 143,985 hectolitres in 1914 to a maximum of 285,357 hectolitres in 1910. Despite the commercial context dominated by significant yearly variations, during the last decade of the nineteenth century, wine exports to Argentina accounted for about 10–15 per cent of total exports, while in the period 1905–1913, they reached 20–26 per cent. At the same time, Italian wine exported to Argentina had a value between 7–9 million and 16–18 million of Italian lire depending on the year: it represented between 19 and 27 per cent of the total amount of exports. Moreover, the Argentine market continued to push Italian winemakers towards wines of higher quality and higher added value. As in the case of the United States (Vaquero Piñeiro and Maffi 2018), this favoured the most dynamic Italian wineries giving them many opportunities to expand their sales.

However, Italian wines had to face the progressive creation of an Argentine wine sector which began to cover a part of the demand traditionally satisfied by European wines. This process of substitution essentially concerned the segment of common wines requiring less technical and professional skills for the winemaking and also easy to market. In 1899, in the provinces of Mendoza and San Juan, the area dedicated to viticulture was 39,116 hectares, with a wine production of over 2.8 million hectolitres: in 1919, vines occupied 114,000 hectares and wine production exceeded 4 million hectolitres (Anderson et al. 2017). Besides, in the early twentieth century, there were 3409 wineries (Girola 1906, 75). As a result of the growth in local oenology supported by a protectionist policy allowing the sale of "petiot" and other sophisticated drinks, from 1883 to 1913 the imports of beverages, above all wine, decreased from 11.6 to 3.3 per cent of the total value of imports (Rayes 2018; Berlinski 2003). Some dynamic wineries were born in the city of Mendoza, in many cases founded by entrepreneurs of Italian origin (Stein and Mateu 2018, 291–322).

Besides wine, other Italian drinks were sent to Argentina: Italian sparkling wines never really threatened the position of French champagne, but sweet wines deserved special mention for their success. In opposite, even if some liqueur wines, primarily Jerez and Porto, were greatly consumed in Buenos Aires, the Marsala failed to fit into the Argentine market (Pedrocco 1994, 329–333), demonstrating that "migrants and national affinity" did not always offer the expected results: Italian immigrants did not consume it because it was considered a "drink of too low a class". Italian vermouth had a great success, especially that "of Torino" that many winemakers tried to copy. It enjoyed enormous favour, primarily the brand Cinzano (Mainardi and Berta 2018, 78–82). In 1887–1888, the Italian vermouth imported into Argentina amounted to about 30,000 litres and 700,000 bottles: in 1905, the Cinzano house alone exported more than 200,000 cases. Other well-known brands were Cora, Carpano, Gancia, Calissano, Ballor, and Isabella (Chiaromonte 1906). Average prices ranged from 19 *pesos nacionales* for a case of Carpano to 14 p/n for Cora vermouth. The consumption of vermouth in Argentina was very high and pushed the production of vermouth in Argentina: it went from about 600,000 litres in 1899 to over 2,285,737 litres in 1905. An Italian company that decided to produce vermouth in Argentina was Martini & Rossi: it was sold at 15–17 p/n per case, a very competitive price compared to imported vermouth. Finally, in the late 1890s, other Italian alcoholic beverages exported were, even if in small quantities, cognac, grappa and bitters (Chiaromonte 1906, 17).

After the First World War and in the first half of the 1920s, the international wine trade strongly declined from 3 to 5 million hectolitres. The main importing market was France, whose imports reached 9 million hectolitres. The United States, which before the war imported between 2 and 2.5 million hectolitres, banned alcoholic beverages, thus closing one of the main outlets for export markets. Germany, which during the Belle Epoque normally imported more than one million hectolitres, greatly reduced its consumption because of the economic difficulties. Russia almost suspended imports after the Soviet Revolution, and even the states emerging from the break-up of the Habsburg Empire reduced their wine imports. The main exporting countries, after Algeria and Spain, were France and Italy, which exported between 1 and 2 million hectolitres. Before the war, the value of wine exports approached 200 million lire for France and Algeria, 100 for Spain, and 70 for Italy (Mortara 1925): in 1923–1924 the value of exports was 275 million lire only.

Concerning the type of wine exported, the main Italian wine was vermouth (59,700 hectolitres, equal to 40 million lire) followed by bottled wines (13,200 hl, 7 million lire) and bottled sparkling wines (4700 hectolitres, 7 million lire). Main sales concerned wines in flasks, kegs, and demijohns: 1.6 million hectolitres and 216.5 million. No relevant organisation managed these exports: the only exceptions referred to vermouth and some table wine in flasks. Wine progressively lost weight among the main Italian products exported: at the end of the nineteenth century it was in the third place, it then dropped to the bottom and disappeared completely in the years 1925–1929 (Federico et al. 2011, 26).

In 1925, Italian wine exports reached 1.3 million hectolitres, a figure that returned to 1936, when the most depressive years of the first half of the 1930s finished. Most relevant countries importing Italian wines were Switzerland (in some years, accounted for half of the exports of Italian wine in casks, demijohns and tank wagons), Argentina, and Brazil (they were in the lead for wine in bottles or flasks). It is true that statistics do not always distinguish between table wine exports and special wines or vermouths. In the 1930s, exports to the African colonies, Libya and mainly Eritrea, were relevant for all bottled and flasks of wine.

Observing the trend in exports (Fig. 4.5), there was no significant change between the end of the Second World War and the early 1970s. Italian wine exports concerning bulk and bottled wine represented a very small percentage of the total wine produced. After the Second World War, the Italian wine sector was not able to have success in international markets (González Esteban et al. 2016; Corsi et al. 2018). In the 1950s, Italian wines sold in foreign markets remained stable at around 1.5 million hectolitres, equivalent to around 1.5–2 per cent of all wine produced. The main markets were Switzerland and Germany, which absorbed almost half of all Italian wine exports. In the early 1960s, exports reached 2 million hectolitres and, in addition to Germany and Switzerland, Austria and the United States gained positions among importer countries.

The real turning point arrived after 1 July 1968, when the European Common Market was definitively opened with the abolition of the last customs barriers. From 1970 to 1972, exports rose from 4.8 million to 13.3 million hectolitres (+177 per cent): southern regions exported large quantities of bulk wine as in the nineteenth century (Fournier 2021). The demand for Italian wine in particular increased in France: it reached 7.8 million hectolitres in 1972, partly to compensate for a poor grape harvest due to drought. So, Italian wine exports exceeded 3.5 million hectolitres and reached the levels of 1887.

The positive trend continued in the following years: exports grew up, reaching 18 million hectolitres in 1979, with the French market dominating (8 million hl). Sales also increased in Germany (3–4 million hl), and in the United States (2 million hl). As total annual wine production in the 1970s was 60–70 million hectolitres, exports accounted for around 20 per cent. This was mainly bulk wine, sent by tanker truck: it was purchased by some large French wine companies. The French viticulturists, losing their guarantees and protections, contrasted the arrival of the low-quality Italian wine, which were sold at very competitive prices. So, in the 1970s Italy was one of the main suppliers of blended wine, as it happened before the customs war of the 1880s. Compared to domestic production, exports jumped to 22 per cent: many poor-quality wines, that previously remained in the internal market, were sold in the international trade.

The favourable market trend allowed Italian wine sector to earn the necessary for new investments improving the quality and in particular favoured wines with a designation of origin. From 1975 to 1985, total exports remained around 20 per cent of the total production, but exports of bottled wine on a bulk basis rose from 24 to 66 per cent and of the total sold in foreign markets from 20 to 40 per cent. Sales in particular increased in Germany and United States.

The highest point of Italian wine production was reached in 1983 with 83 million hectolitres, but in 1986, the methanol scandal broke out, demonstrating the presence of fraud and dangerous manipulation. Besides, the designation of origin was not fully applied, and in 1984 a law taxed methanol, making it cheaper than sugar, the use of which by mixing it with wine had been banned in Italy since 1965. Exports fell to 11 million and per capita consumption fell drastically from 68 to 35 litres: however, production rates remained high, 60–70 million hectolitres.

The situation changed starting in the mid-1990s: 1995 exports (18,907 million hectolitres) accounted for 29 per cent of total production. There was a strong reduction in wines with an alcohol content above 13 degrees in favour of "still" wines below 13 degrees, which rose to 78 per cent; then came sparkling wines (12 per cent), semi-sparkling wines (5 per cent), and musts (4 per cent). Full-bodied wines fell to just 2 per cent. The best performances for PDO wines were in Germany and the United States: together with France and Great Britain, they absorbed 72 per cent of Italian wines. As a result of the drop in production to around 50 million hectolitres in 2000, exports (18,500 million hl) rose to 34 per cent, a percentage that continued to grow in the early years of the new millennium, reaching 58 per cent in 2012 (22,355 million hectolitres). Italy became a net importer of bulk wine from other countries, with Spain in the lead.

The success of the Italian wine in the international markets continued during the period 1995–2005, when international trade in wine practically doubled in value, going from 10.2 billion dollars in 1995 to almost 20 billion dollars in 2005. Of this total, approximately 54 per cent was destined for European markets. The quantities traded increased by "only" 48 per cent over the same period highlighting the significant increase in the export price per unit of product. The increase in world import-export registered a decisive surge from 2002 onwards, with growth rates always well above 10 per cent per year, and a peak in 2003 when +21.9 per cent was recorded compared to the previous period. From 2004 to 2005, the quantities exported did not change.

This rapid evolution of the wine trade reflected the renewed needs of a market where the methods, frequency, and places of consumption of the product were changing, and there existed a clear segmentation of consumers. In order to make the most of their production structure, the major competitors in the sector shifted an ever-increasing share of their production from the domestic market to exports, with the aim of compensating for the simultaneous growth in imports of foreign products, and above all to conquer the wider profit margins that international markets provided (UnionCamere 2007).

In the European context, Italy represented a secondary import market: Italian wine imports were worth around 260 million euros in 2005. A modest figure if compared to the 9.3 billion euros of European imports. More than half of Italian imports were sparkling wines, even though the incidence of this category was declining. In opposite, the contribution of wines remained constant.

Purchases of bulk wine from abroad grew more than the average: Californian wine producers exported bulk wine to Italy to be packaged and sold in Northern European markets. This practice, existing in the early twentieth century, brought the United States to first place among the suppliers of bulk wine, rising from 12.9 in 2004 to 51.1 per cent in 2005 and contributing to the growth of the weight of bulk wine on the total of wine imports, It also brought the United States to second place on the total of Italian wine imports (15.6 per cent in 2005) (4.10). France remained the largest exporter of wine to Italy with 62.9 per cent of Italian imports, thanks mainly to the sparkling wine sector (champagne) where it holds 95.8 per cent of the market. In parallel with the reduction of the relevance of this sector, France lost weight on the Italian market (in 1998 its market share was 81.9 per cent): however, it maintained its leadership role in the bottled wine sector.

Performances of the major European producers varied substantially. Spain and Italy maintained their market shares almost unchanged (around 9.6 and 18.8 per cent respectively) (Fig. 4.5), thanks to a trend that was substantially in line with the evolution of world trade (Spahni 2000). France most suffered from competition of wines coming from Austral

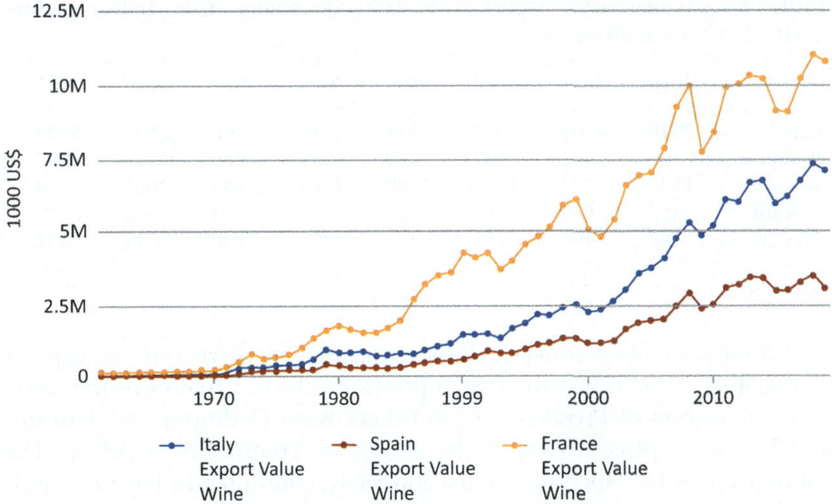

Fig. 4.6 Italy, Spain, and France wine. Value (1961–2019). (Source: FAOSTAT)

hemisphere: its share of international exports fell from 44.6 in 1995 to 35.5 per cent in 2005, with a worrying loss of competitiveness everywhere. In 2005, with an export value of over 3 billion dollars and a market share of around 18 per cent, Italy confirmed its position as the world's second largest exporter (6.9 billion dollars) (Fig. 4.6). In a context where new competitors emerged, Italian exports were in line with the development of world wine trade (+90 per cent from 1995 to 2004 and +3.4 from 2004 to 2005).

In the period 1998–2005, about 9 per cent of Italian exports shifted from bulk wines to bottled wines, which in 2005 accounted for 83.8 per cent of the total value exported. The United States became the leading market for Italian wines (26.2 per cent of total wine exports in 2005) followed by Germany (23.4 per cent). There were profound differences between Italy's two main markets: the United States bought from Italy almost exclusively bottled wines (29.2 per cent of Italian wines in this segment) and sparkling wines (17.2 per cent). Germany purchased considerable quantities of bulk wine (39.7 per cent); bulk wine exports continued to be supported by Switzerland (14.8 per cent). The other important markets for Italian exports were, in order, the United Kingdom (12.5 per cent), Canada (5 per cent), and Japan (3.2 per cent).

Table 4.4 Italian wine export with data concerning main Italian regions (2010–2017) (million euros)

	2010	*2011*	*2012*	*2013*	*2014*	*2015*	*2016*	*2017*
Italy	3918	4405	4696	5042	5115	5390	5628	5990
Piedmont	775	891	887	969	985	965	924	987
Veneto	1158	1332	1444	1588	1671	1834	2001	2129
Tuscany	587	659	703	748	762	902	918	937
Trentino	407	444	451	477	492	500	511	532

Source: ISTAT

Strong disparities concerned the contribution of regions to exports (Table 4.4). Veneto was the first, a position achieved thanks to the international success of Prosecco; behind there were Piedmont and Tuscany, and in fourth place appeared the emerging Trentino-Alto Adige. The southern regions, especially Apulia and Sicily, continued to have a considerable relevance.

Before Covid-19 and its effects on world markets, a substantial share of Italian wines was exported, for a value of 5.89 billion in 2018 (almost 8 times that of 1990) (Mediobanca 2020). This depended on the increase in both sparkling wines (+4.5 per cent) and still wines (+2.7 per cent). Exports grew in Europe (+1.9 per cent, with a 60 per cent share), America (+4.7 per cent with a 31 per cent share), Asia (+7.1 per cent with 7.3 per cent), and even Oceania and Africa, although together absorbing just over 1 per cent of sales, were both expanding by 4.1 and 15.4 per cent respectively. Most important markets for Italian wine increased (excluding the United Kingdom, which contracted by 5 per cent): +5.4 per cent for the United States, +2.5 per cent for Germany, +1.6 per cent for Switzerland, +2.6 per cent for Canada, and +6.5 per cent for France. The top six destinations (all over 200 million) absorbed 66.9 per cent of the value of Italian wine in foreign market (only the top three accounted for 52.5 per cent). In addition to Japan (+13.1 per cent), the countries where Italian wine exports exceeded 50 million euros and were growing at double figures were Russia (+21.1 per cent) and Poland (+14.1 per cent). The Netherlands (+8.8 per cent) and China (+5.4 per cent) also performed well.

Most Italian wine exports went to European countries (51.9 per cent of foreign sales in 2019, up 3.4 per cent on 2018 when they weighed 52.7 per cent). North America was the second largest area, expanding slightly to 31.7 per cent of the total (from 31.2 per cent, up 6.5 per cent). Africa, the Middle East and Not-EU European countries represented 9.7 per cent (up 10.9 per cent), while Far East markets and Latin America put in evidence diversified trends (+3.6 and -8.2 per cent respectively) and remained marginal (5.3 and 1.4 per cent of the total).

4.3 Domestic Consumption

In contrast to the myth of the wine consumed by peasant families, the sources indicate a reality dominated by low wine consumption in the Italian countryside at the end of the nineteenth century. So, in the Italian countryside the wine was produced but not drunk. Besides, when Italian peasants consumed wine, they drank "acquarello", "acquaticcio" or "vinello", that is their wines were diluted in water. Even in Piedmont, except in the "winemaking villages" where it was possible to dine on good wine, the norm was to eat meals with little wine or with "posca" or "pichetta", that is a very light wine, almost like water (Atti della Giunta 1883). In Italy there was the same situation existing in France, where piquettes were a great success (Simpson 2011), and in Argentina with petiot: peasants consumed wine-based drinks laying on the fine line between falsification and adulteration.

The only exceptions were represented by a few days of festivity and during the heavy summer work, when the arrival of peasant workers, relatives and neighbours to complete the grain harvest forced landowners (or tenants) to offer substantial lunches as a sign of welcome and physical comfort (Betri 1998). For the rest of the year, deprivation or assiduous frequentation of taverns prevailed. Italian peasants consumed few pure wine in favour of very light beverages made from the last pressing of the grapes dominated: however some of them visited taverns where they drank wines together with workers and artisans (Sorcinelli 1999). So, there was a domestic consumption marked by moderation, including at most a few small wines or glasses of pure wine (i.e. the actual peasants' wine self-consumption was less relevant), and there also existed a consumption outside the home, in public places (Barberis 1983).

The wine consumption inside and outside the family makes statistical calculation very problematic (Simpson 2011) and explains discrepancies between the official estimates of widespread wine consumption and the numerous documents indicating that working classes consumed few wine, and often did not include it in their meals (Somogyi 1960, 1975). The rural lower classes were in fact humble people of modest consumption, attentive to work and social respect, in contrast to the popular masses in the cities, more liberal and ready to question the existing social balances: this explains why all European and US governments were against the consumption of alcohol (Cavazza 2004).

For the period 1861–1900, the average annual consumption of wine in Italy was around 84 litres per capita (Barberi 1961; Circolo enofilo italiano 1892, 1893). In the same period, France, with 153 litres per capita, had the world's highest per capita consumption: the consumption in fact doubled (Pinilla and Ayuda 2007, 11). In France as well as in Italy, these data considered both low-quality wine mixed with water and good imported wines.

In the Italian larger cities, the average annual consumption of wine rose to 126 litres, while in the smaller ones it fell to 64 (Quirino 1991; Zamagni 1998). The consumption was higher in the Northern urban centres, but a correct evaluation is difficult because some wages (in particular in the countryside) were partly paid by delivering quantities of wine. The internal wine market was mainly fuelled by products from the southern regions and Piedmont: they were sold in the major centres of consumption where, as in the cases of Rome, Turin, Florence and Verona, the quantity available for consumption per inhabitant at up to 150–200 litres (Friz 1980; Zaninelli 1974; Balestrieri and Gini 1924; Trentin 1895, 1903). In the new millennium the consumption increased: in 1900 the wine in the internal market was 34.8 million hectolitres for 32.9 million inhabitants, therefore there were an availability of more than 100 litres for each person including children; in 1916, during the war, the domestic availability amounted at 44 million hectolitres, this means an individual consumption of 128 litres (Mondini 1916).

In Milan, at the end of the nineteenth century there was an increase of 27.5 per cent in the number of taverns and inns, while the increase in the number of restaurants, breweries, wine bars, and cafés was as high as 146.77 per cent (Zamagni 1998). These data testified the widespread habit of the working class, but also of rural workers, of spending a large part of their time drinking wine in meeting places, a practice which, as well as deserving of moralising campaigns, stimulated investigations and medical complaints on the physical and psychological ills connected with the excessive consumption of alcoholic beverages, many of them consisting of wines adulterated with substances of dubious origin (Verga 1873; Ceccarelli 1874). Debates on the production and consumption of wine illustrated two main opinions and proposals: to strongly limit the consumption as the wine was a source of vices and behaviour outside civilised life; to increase the domestic market to better balance supply and demand in the very frequent case of very good grape harvests and related falling prices. In the early years of the twentieth century, winemakers called for a public intervention in times of excessive production, for public economic facilities concerning the rail transport of southern wines to the urban centres of the north, for public loans allowing the greasing of cellars or the purchase of wine pots, for the fiscal facilities favouring the creation of wine cooperatives (Corbino 1938).

Almost all Italian wines were involved in the internal market: liqueur wines and special wines were actively traded, including Marsala, vermouth, moscato, and spumante. The wine sector was subordinate to the trend of the grape harvest setting the prices on the market. There was a lack of balancing mechanisms: exports were inadequate to compensate for a growing domestic supply. Besides, popular consumption in the towns and countryside was dominated by "vinelli" and watered-down wines: there was also the practice of mixing "vinelli". This aspect, which official data could not verify, obviously increased the difference between the wine production and the actual consumption. This also pushed winemakers to ask for tax concessions and the authorisation to take the surplus wine to distilleries producing liqueurs, vinegar, cognac, bitters, digestives, and spirits made from wine.

Concerning prices, according to official statistics in Italy, as in France and Spain (Planas 2016; Barrio 2010), after a long period of fluctuation from 1861 to 1885, from 1886 prices progressively fell and the depressive trend continued until the early twentieth century. Doubts about prices were widespread due to the high consumption of "vinelli" escaping statistical surveys. In the period 1875–1894 between the large consumption cities of the north and the production areas of the south, there was a difference in price of 178 per cent: in Milan, the main town of Lombardy, the average price of a hectolitre of wine was 61.08 lire, while it was 21.96 lire in Catania, one of the main Sicilian markets.

The commercial activity was dominated by large merchants hoarding wine in the terroirs and then sending it to the consumption areas according to the trend in demand (Mondini 1916, 86–88). These big wine merchants imposed the price: most winemakers in fact had a low contractual power. An evident clash existed between these two groups, and this favoured the creation of winemakers' associations reducing the excessive fragmentation in terms of supply and distribution.

In 1884, the "Società generale di produttori di vino" (General Society of Wine Producers) was founded; in 1895, the "Unione vini italiani" (Union of Italian Wines) was created: finally, in 1917 the "Federvini" (Federation of Italian producers, exporters and importers of wine, spirits, liqueurs, syrups, vinegars, and similar) was established. At the same time, in some regions, such as Lombardy, associations of wine sellers and merchants came into operation (Unione Italiana Vini 1925). The main aim of these association bodies was to obtain by public administration measures defending the wine sector and giving the internal wine market more stability.

In 1927, the "salone per le contrattazioni vinicole" (literally a room for wine negotiations) was founded in Milan and it became a section of the Milanese stock exchange in 1929. Before the 1920s, some wine exchanges or trading centres existed in Northern Italy, in Casalmonferrato, Verona, and Alessandria: they represented a reference for prices. Even if local prices obviously fluctuated, for the large wine operators Milan became a relevant meeting point between the large southern merchants and the capillary distribution chain existing in the northern cities. It facilitated relations between wine producers, shopkeepers, and consumers by providing information on prices and the situation in individual cities, without neglecting the provision of analysis, advice, and technical consultancy. Other associations, such as the wine cooperatives, winemakers unions, and the experimental wine cellars of Barletta and Riposto (in Southern Italy) performed a similar function: they were places where operators could meet and do business. The Milan stock exchange also attempted to give Italian wine a uniformity favouring its commerce: the wines brought to the Milan wine exchange in fact had to be healthy and without legal or commercial defects. Wines with extraneous odours and flavours, altered colour and clarity were not permitted, nor could wines with too little alcohol be traded. There were two basic wines for negotiation: red wine for meals of 10 degrees and red wine for blends of 13.5 degrees. Finally, in order to be admitted to the wine exchange, operators had to have at least 300 hectolitres of wine to sell.

Falling prices, the closure of national and international markets and falling family incomes characterised the wine sector between the two world wars. In the 1920s, per capita consumption was 80–95 litres: too little for a country with a wine production of over 40 million hectolitres and without a foreign market to sell the surpluses. So, an abundant grapes harvest was enough to create a crisis for winemakers because of the fall in prices

(Marescalchi 1925; Grando and Volpi 2014). In gold lire, not in overly undervalued paper lire, the value of Italian wine and viticulture was lower than during the Belle Epoque. The Italian wine sector in fact faced the sharp increase concerning the cost of labour and the price of the products used to cultivate vines. This was accompanied by a contraction in wine consumption. The cost of wine, reduced to gold lire, was 1.30–1.80 times that of the pre-war period, while the selling price was only 0.5–0.7 times.

During this negative trend, in a climate of increasing exaltation of patriotic values, the grape festival was instituted (Cavazza 1997). It consisted of picturesque ceremonies, animated by parades of allegorical floats, followed by solemn awards and blessings. Appropriately reported by the press, which was essential in documenting the whole celebration with a rich photographic apparatus and articles full of patriotic rhetoric, the ceremonies took place "in an atmosphere of vibrant enthusiasm" and were mainly attended by organised groups of rural housewives and young girls dressed in picturesque regional costumes (De Grazia 1992; Salvatici 1999; Willson 2002). So, for a few days, the population as a whole, overcoming social and class barriers, was given the task of filling the streets of the villages to take part in a festival to celebrate the consumption of a typically national fruit. Like many other products whose consumption found its reason for being in the autarchic model imposed by the fascist regime, grapes too (called Regina or Italia) had to contribute fully to the national development (Scarpellini 2008).

The grape festival had to promote rational winegrowing, especially in those regions where, while there was an abundant production of wine grapes, the production of grapes for direct consumption was very low (not even 2 per cent). However, the table grapes brought into the city remained unsold because they cost too much and were actually considered alien to the population's eating habits. As it continued to happen in the 1930s, despite the extensive deployment of propaganda media, Italian grapes were considered to be used to make wine, not to be eaten as fruit.

Other aspects emphasised in the press as an example of rational agriculture in keeping with the practical spirit of Fascism were represented by the experiments in recovering marc and using it to make a wide range of products (alcohol, tartaric acid, cake, fertilisers, etc.).

In an attempt to portray the image of a modern country, vineyards, wine cellars and popular taverns, appropriately reinterpreted in terms of the genuine values of the Italian people, started to appear in magazines. Fascist regime, copying what happened in other European countries such as Germany, France, or Switzerland, had to know how to create a tourist and leisure offer that included visits to wineries and wine-growing landscapes. In order to attract the interest of those who had the economic possibilities to "go on holiday" in the lake riviera, in spa and mountain (Cosmo 1939; Battilani 2009; Berrino 2011), the restaurants of the best hotels were equipped with picturesque taverns where tourists could drink and taste some popular flavours in an informal and convivial manner (Le taverne 1939). Similarly, it was necessary to learn how to decorate and embellish wine bars in order to make them pleasant places, leaving behind the idea that wine was consumed in dark, unpleasant-smelling establishments. In a climate of rediscovery of territories and gastronomic itineraries, which had as its corollary the creation of a positive image of wine, the first gastronomic guide of the Italian Touring Club was published in 1931: it aimed at highlighting "what one eats and how one drinks", above all with the tastes of those who, for leisure and entertainment, travelled by car along the Italian roads. The guide indicated more than 150 types of wine which were very well known, or were the expression of a typically local dimension. The Touring guide, although a text intended for tourists, provided for each wine interesting information about the terroirs. Besides, the wines were named using extensive territorial districts (Chianti) or municipalities (Rufina, Pomino, Carmignano, Barolo). So, in the early 1930s, it became relevant the weight of "places" in giving individual wines a specific reputation: it predominated the idea that territories gave wines their character and reputation (Livat 2019). The 1963 law on designations of origin de facto adopted this criterion giving the classification a distinct land-based sense (Navarini 2015, 86–92).

After the Second World War, the wine sector again presented strong imbalances on the supply side, which was far in excess of consumption: prices declined. The neo-realism cinema illustrated the social difference existing between a small minority who in public places boasted of being able to order a famous wine and the working classes for whom wine in a bottle and with a label represented a real luxury, an excessive expense (Gallingani 1998; Attorre 2009; Coviello 2012; Lepratto 2016). A parliamentary enquiry into poverty conducted in 1952 indicated that more than 3.3 million families never consumed wine because they had no money (Braghin 1978).

In the 1950s, the only wine that continued to enjoy a good reputation was Chianti but, in the absence of precise legislation on production areas, any wine bottled in the classic stuffed flask bore the designation. This was the result of a precise legislative orientation that did not favour best winemakers. Exports did not even reach one million hectolitres. At the same time, some relevant changes took place: the reduction of wine consumption in taverns was linked to the diffusion of bars frequented by the new generations attracted by other tastes and lifestyles (Nowak 2019). So, politicians criticised Coca-Cola consumption and indicated the wine as a symbol of "an endangered cultural Italianness" (Bernardi 2020, 100–102): they also called for protectionist measures defending such a national product as wine. The wine was used as a tool to obtain political consensus, especially in the countryside. Face to first step of the European integration, peasants' trade unions as Coldiretti gave a negative evaluation and highlighted the problem of Italian agriculture in a more competitive market. The modern cities with their progressist habits were considered a threat to the integrity of tradition rural society representing a fundamental support for the governments contrasting the expansion of parties linked to the Soviet Union (Barberis 1999). Italian wine and the old village taverns, in this climate of fervent nationalism transferred to food, constituted a real cultural dam, even if the quality was very low and most of wine was consumed mixed with water to make it drinkable and to spend less. In Italy, 60 per cent of the population did not consume wine outside home (Luzzatto Fegiz 1952).

The household spending remained "extremely simple", dominated by essential foods (Di Bernardi 2015; Helstosky 2018). Conservation and saving predominated household strategies: the aim was the sustenance of the family (Somogyi 1960). So, wine consumption was considered a marginal expense. In northern regions wine came after bread and milk, while in southern regions it came after bread, soup, and vegetables. Changes essentially depended on best entrepreneurs' commercial strategy modifying the culture of food consumption, including wine (Cavazza 2010). In 1950, the Piedmontese company Giordano Vini began to sell by mail order: it was the first step of changes allowing, some years after, to sell wine bottles on supermarket shelves, a choice creating scandal among "wine purists" (Scarpellini 2008). Besides, the arrival of the television and its ability in persuading consumers through advertising progressively modified the best wine companies' marketing strategies and allowed them to enlarge their sales (Capuzzo 2015).

The Piedmontese journalist Mario Soldati (1906–1999) perfectly grasped the transitional phase that the domestic wine market in Italy was about to undergo, making the traditional drink into a sort of kaleidoscope through which to observe the consolidation of behavioural and consumption models typical of modern and industrialised societies. He thought that wine consumption had to prove a bulwark in defence of the peasant authenticity that was about to disappear (Soldati 2011). When choosing and consuming wine, Italian consumers performed a kind of cultural resistance, whether at home or in a restaurant. They had to remove their idea that bottles of wine with labels were almost always bad: most consumers in fact thought that Italian wines could deserve the adjective "exquisite" only if they remained faithful to their productive tradition, abandoning any pretence of becoming "fine". At the same time, he criticised the massification and standardisation of consumption, which forgot the authentic flavours and aromas of Italian wines. The genuine wine was a drink full of authentic values, and it was in opposition to consumerism, industrialisation, planning, and new fashion. Soldati's theses represented the clash between the traditional wine and the modern one which was arriving on the market (Image 4.2).

Mario Soldati, in his televised account of the Italian regions, presented wines and typical gastronomic products to the general public, Soldati demonstrated his ability in rediscovering territories (Capatti 2015). He used the mass media in order to carry out a capillary marketing and

Image 4.2 Grape harvest in the Valtellina valley (1960s). (Source: Archivio Storico Unione Italiana Vini—Milano)

advertising operation (Manetti et al. 2006; Pastore 2004). Even if he desire to combat consumerism as the cause of the loss of flavours and genuine products, he advertised and disseminated good Italian wines in bottles (Petrini and Padovani 2017).

With television reports during the economic miracle, the Italian consumers started to know good wines, which went from being local to national and within everyone's wages (Vecchi 2017). Luigi Veronelli, who was also a witness to the contradictions arising from a peasant society undergoing the trials of modernity, started to explain the relevance of traditional wines and their flavour and taste. His famous statement was: "the last of the artisan wines will always be better than the first of the industrial wines because it will have a soul". It was difficult to reconcile the defence of typicality with the need for consumers to learn more about the characteristics of quality wines. They needed to be educated and guided in their wine choices, in particular the working classes which earned more and opened up to a different kind of wine consumption, seen as a symbol of a new lifestyle of pleasure and leisure. Magazines gave information on which glasses were a must in a modern home: so, without becoming experts, consumers were encouraged to learn about the characteristics of wines to pair with food ("La cucina italiana", year V, July, 1956, 478).

Popular magazines contained texts and advertisements illustrating the new status symbols of domestic well-being, from electrical appliances to Formica furniture, from plastic containers to pantries full of tinned preserves bought at the supermarket. In this context, the wine slowly evolved from a food commodity to a leisurely blessing. These new objects of desire also included a small wine cellar containing a selection of bottled wines: Italian consumers in fact allocated a small part of their income to the purchase of some gastronomic pleasure or fine wine (Gabrielli 2011).

At the end of the 1950s and before the approval of the law on designation of origin, the Italian wine market was marked by tensions. The controversy that accompanied the approval in 1956 of the law making it compulsory to bottle vermouth was significant. For many, it had to be sold in bulk as sales in bottles increased prices due to taxation. However, bottled wines were sold more easily, as the more modern shops in Paris demonstrated. Another innovation appearing in the mid-1950s was represented by plastic containers, practical and convenient for transporting bottles. When in 1956 the US government set up a test supermarket in Rome to demonstrate to the Italian authorities the advantages of the American self-service method in food distribution (De Grazia 2005), it was evident that new selling ways for wine were opened. Besides, most consumers did not want to buy wines of little quality, at permanent risk of fraud, adulteration, and tampering.

However, political institutions were not able to understand needs related to the modernisation: the umpteenth failure in 1957 of the project of a new law defending typical or fine wines showed that most winemakers did not want to (or could not) renew the Italian wine sector. So, the real turning point was represented by the progressive creation of the European Common Market and the new Common Agricultural Policy: they in fact obliged to create a strict wine register and to start the renewal of old wineries, and it was necessary a relevant public aid to finance winemakers.

However, even if the economic miracle allowed to increase the quality of life, the actual growth of wine consumption did not increase proportionally. Between the 1960s and 1970s in Italy the average per capita consumption was around 90 litres, a very similar quantity to the end of the nineteenth century. This means that Italian winemakers had to reduce their production and improve the wine quality to increase the sales on foreign markets.

Even if in some terroirs the renewal and the related success of the wines arrived during the 1970s, the real affirmation of the Italian wine sector started in the second half of the 1980s, after the scandal of wine adulterated with methanol and the arrival on the European market of the competitive Greek, Portuguese and Spanish wines as well as the new Californian wines which outclassed the full-bodied French ones.

It was necessary to produce (and consume) less but better (Navarini 2015) and in fact, during the last decades of the twentieth century, wine consumption gradually decreased (Zamagni 1998). While in the 1960s per capita wine consumption exceeded 100 litres per year, in the 1990s it fell to less than 54 arriving at the end of the century to just 47. At the beginning of the new millennium per capita wine consumption in Italy was just 34–38 litres.

At the same time the Italian wine built a new positive image presenting a wide range of products, capable of intercepting the demands of an increasingly young and cultured clientele. Wine tourism also played a relevant role. So prices, which started to increase in the second half of the 1970s and in the 1980s because of the inflation more than the growth of the demand, were finally able to increase because the product interested consumers who were disposed to pay more to drink better. The impetuous growth reached a peak in 2004: after a slight contraction between 2005 and 2009 (-7 per cent), the positive trend continued in the following years, reaching a new peak in 2014 (+26 per cent) (Fig. 4.7).

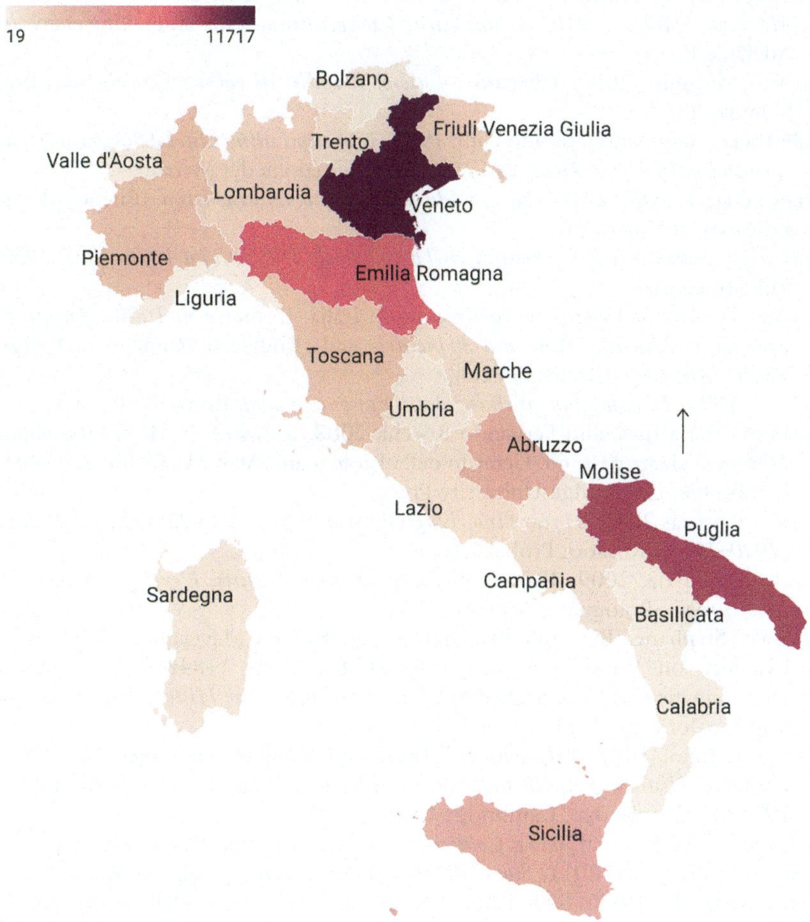

Fig. 4.7 Wine production divided by regions (2020) (thousands of hectolitres). (Source: ISTAT)

References

Anderson, Kym, Signe Nelgen, and Vicente Pinilla, eds. 2017. *Global Wine Markets, 1860 to 2016: A Statistical Compendium*. Adelaide: University of Adelaide Press.

Attorre, Antonio. 2009. *Chateau Lumiére. Brindisi ed ebbrezze al cinema*. Bra: Slow Food.

Balestrieri, Mario, and Corrado Gini. 1924. *I consumi alimentari della popolazione italiana dal 1910 al 1921*. Padova: Amministrazione del Metron.

Balmaceda, Daniel. 2016. *La comida en la historia argentina*. Buenos Aires: Editorial Sudamericana.

Barberi, Benedetto. 1961. *I consumi nel primo secolo dell'Unità d'Italia 1861–1960*. Milano: Giuffrè.

Barberis, Corrado. 1983. L'autoconsumo in Italia. In *Storia di Italia. Annali 6. Economia naturale, economia monetaria*, ed. Ruggiero Romano and Ugo Tucci, 746–774. Torino: Einaudi.

———. 1999. *Le campagne italiane dall'Ottocento a oggi*. Roma-Bari: Laterza.

Barbero, María Inés, and Fernando Rocchi. 2003. Industry. In *A New Economic History of Argentina*, ed. Gerardo della Paolera and Alan M. Taylor, 261–294. Cambridge: Cambridge University Press.

Barrio, Patricia. 2010. *Hacer vino. Empresarios vitivinícolas y Estado en Mendoza (1900–1912)*. Rosario: Prohistoria.

Battilani, Patrizia. 2009. *Vacanze di pochi, vacanze di tutti. L'evoluzione del turismo europeo*. Bologna: Il Mulino.

Becuwe, Stéphane, Bertrand Blancheton, and Samuel Maveyraud. 2022. New Evidence on Wine in French International Trade (1848–1913): Import Discrimination as Export Quality Promotion. *Economic History Review*: 1–28. https://doi.org/10.1111/ehr.13156.

Berlinski, Julio. 2003. International Trade and Commercial Policy. In *A New Economic History of Argentina*, ed. Gerardo della Paolera and Alan M. Taylor, 197–232. Cambridge: Cambridge University Press.

Bernardi, Emanuele. 2020. *La Coldiretti e la storia d'Italia*. Roma: Donzelli.

Berrino, Annunziata. 2011. *Storia del turismo in Italia*. Bologna: Il Mulino.

Betri, Maria L. 1998. L'alimentazione popolare nell'Italia dell'Ottocento. In *Storia d'Italia. Annali 13, L'alimentazione*, ed. Alberto Capatti, Alberto De Bernardi, and Angelo Varni, 7–22. Torino: Einaudi.

Bonardi, Luca. 2014. Spazio e produzione vitivinicola in Italia dall'Unità a oggi. Tendenze e tappe principali. Territoires du vin [Online], 6 | 2014. Online since 1 March 2014, connection on 1 February 2021. http://preo.u-bourgogne.fr/territoiresduvin/index.php?id=821.

Braghin, Paolo. 1978. *Inchiesta sulla miseria in Italia*. Torino: Einaudi.

Brassley, Paul. 1997. *Agricultural Economics and the CAP: An Introduction*. Oxford: Blackwell.

Brassley, Paul, Carin Martiin, and Juan Pan-Montojo. 2016. European Agriculture, 1945–1960: An Introduction. In *Agriculture in Capitalist Europe, 1945–1960: From Food Shortages to Food*, ed. Carin Martiin, Juan Pan-Montojo, and Paul Brassley, 1–21. New York: Routledge.

Capatti, Alberto. 2015. *Scoperta e invenzione della cucina regionale*. In *L'Italia e le sue regioni*, ed. Mariuccia Salvati and Loredana Sciolla, 409–426. Roma: Istituto dell'Enciclopedia Italiana fondata da Giovanni Treccani.

Capuzzo, Paolo. 2015. Consumi e distribuzione: una storia in cifre. In *L'Italia e le sue regioni*, ed. Mariuccia Salvati and Loredana Sciolla, 513–537. Roma: Istituto dell'Enciclopedia Italiana fondata da Giovanni Treccani.

Casalini, Mario. 1953. *Le cantine sociali cooperative*. Roma: Centro sociali cooperative.

Cavazza, Stefano. 1997. *Piccole patrie. Feste popolari tra regione e nazione durante il fascismo*. Bologna: Il Mulino.

———. 2004. *Dimensione massa. Individui, folle, consumi. 1830–1945*. Bologna: Il Mulino.

———. 2010. Dal consumo desiderato al consumo realizzato: l'avvento della società dei consumi nell'Italia postbellica. In *La rivoluzione dei consumi. Società di massa e benessere in Europa 1945–2000*, ed. Stefano Cavazza and Emanuela Scarpellini, 53–78. Bologna: Il Mulino.

Ceccarelli, Faustino. 1874. *L'ubbriachezza sotto il rapporto igienico, patologico e morale. Letture popolari*. Roma: tip. Poliglotta.

Chevet, Jean-Michel, Eva Fernández, Eric Giraud-Héraud, and Vicente Pinilla. 2018. France. In *Wine Globalization. A New Comparative History*, ed. Kym Anderson and Vicente Pinilla, 55–91. Cambridge: Cambridge University Press.

Chiaromonte, Tommaso. 1906. *Il commercio dei vini nella Repubblica Argentina nel decennio 1895–1904. Istruzioni e consigli agli esportatori italiani*. Roma: Tip. nazionale G. Bertero e C.

Circolo enofilo italiano, ed. 1892. *Annuario generale per la viticoltura e la enologia 1892*. Roma: Circolo enofilo italiano.

———, ed. 1893. *Annuario generale per la viticoltura e la enologia 1893*. Roma: Circolo enofilo italiano.

Corbino, Epicarmo. 1938. *Annali dell'economia italiana 5 (1901–1914)*. Città Di Castello: Soc. An. Tip. "Leonardo da Vinci".

Correa, Carolina, and Matías Wibaux. 2000. Sabores de la pampa. Dieta y hábitos de consumo en la frontera bonaerense. In *Vivir en la frontera. La casa, la dieta, la pulpería, la escuela (1770–1870)*, ed. Carlos A. Mayo, 71–86. Buenos Aires: Editorial Biblos, Historias Americanas.

Corsi, Alessandro, Eugenio Pomarici, and Roberta Sardone. 2018. In "Italy from 1939". In *Wine Globalization. A New Comparative History*, ed. Kym Anderson and Vicente Pinilla, 153–177. New York: Cambridge University Press.

Cortés Conde, Roberto. 1979. *El progreso argentino, 1880–1914*. Buenos Aires: Editorial Sudamericana.

Cosmo, Italo. 1939. I vigneti lungo la 'DeutscheWeinstrasse' e la 'Rue des Caves'. *Enotria* 38: 236–242.

Coviello, Massimiliano. 2012. *Il cinema italiano racconta il vino. Le vigne e le tavole, le produzioni e il consumo consapevole*. Siena: Enoteca Italiana.

D'Onofrio, Federico. 2016. *Observing Agriculture in Early Twentieth-Century Italy. Agricultural Economists and Statistics*. London: Routledge.

Dandolo, Francesco. 2010. *Vigneti fragili. Espansione e crisi della viticoltura nel Mezzogiorno in età liberale*. Napoli: Guida.

De Ferrière Le Vayer, Marc, and Jean-Pierre Williot. 2013. Northern France, 1750–2000. In *The Agro-Food Market. Production, Distribution and Consumption*, ed. Leen Van Molle and Yves Segres, 165–195. Turnhout: Brepols.

De Grazia, Victoria. 1992. *How Fascism Ruled Women. Italy, 1922–1945*. Berkeley: University of California Press.

———. 2005. *Irresistible Empire. America's Advance through 20th-Century Europe*. Cambridge: Cambridge University Press.

della Paolera, Gerardo, and Martín Grandes. 2007. The True Measure of Country Risk: A Primer on the Interrelations between Solvency and the Polity Structure of Emerging Markets, Argentina 1886–1892. In *The Decline of Latin American Economies: Growth, Institutions, and Crises*, ed. Sebastian Edwards, Gerardo Esquivel, and Graciela Márquez, 195–212. Chicago and London: The University of Chicago Press.

Devoto, Fernando. 2001. In Argentina. In *Storia dell'emigrazione italiana. Arrivi. Vol. 2*, ed. Piero Bevilacqua, Andreina De Clementi, and Emilio Franzina, 25–54. Roma: Donzelli.

De Bernardi, Alberto. 2015. Consumi alimentari in Italia: uno specchio del cambiamento. In *L'Italia e le sue regioni: l'età repubblicana. Territori*, ed. Mariuccia Salvati and Loredana Sciolla, 487–509. Roma: Istituto dell'Enciclopedia Italiana fondata da Giovanni Treccani.

Fabiani, Guido. 1987. *L'agricoltura italiana tra sviluppo e crisi (1945–1985)*. Bologna: Il Mulino.

———. 2015. *Agricoltura-mondo. La storia contemporanea e gli scenari futuri*. Roma: Donzelli.

Fauri, Francesca. 2010. *Il Piano Marshall e l'Italia. Bologna*: Il Mulino.

———. 2015. *Storia economica delle migrazioni italiane*. Bologna: Il Mulino.

Fauri, Francesca, and Paolo Tedeschi, eds. 2011. *Novel Outlooks on the Marshall Plan: American Aid and European Re-Industrialization*. Brussels: Peter Lang.

Federico, Giovanni. 1982. Per una valutazione critica delle statistiche della produzione agricola italiana dopo l'Unità (1860–1913). *Società e Storia* 5: 7–130.

———. 1991. Oltre frontiera: l'Italia nel mercato agricolo internazionale. In *Storia dell'agricoltura italiana in età contemporanea. III. Mercati e istituzioni*, ed. Piero Bevilacqua, 189–222. Venezia: Marsilio.

———. 2003. Le nuove stime della produzione agricola italiana, 1860–1910: primi risultati ed implicazioni. *Rivista di Storia Economica* 19: 359–381.

Federico, Giovanni, and Pablo Martinelli. 2018. Italy to 1938. In *Wine Globalization. A New Comparative History*, ed. Kym Anderson and Vicente Pinilla, 130–152. Cambridge: Cambridge University Press.

Federico, Giovanni, Sandra Natoli, Giuseppe Tattara, and Michelangelo Vasta. 2011. *Il commercio estero italiano 1862–1950*. Roma-Bari: Laterza.

Fernández, Eva. 2014. Selling Agricultural Products: Farmers' Co-operatives in Production and Marketing, 1880–1930. *Business History* 56: 547–568.

Ferrer-Alos, Llorenç. 2021. Commercial Structures and the Globalisation of Viticulture in Catalonia: From Brandy to Wine and Cava (Eighteenth–Twentieth Centuries). *The Historical Review/La Revue Historique* 17: 251–304.

Fonseca, Antonio. 1891. *Notizie sull'enologia e sul commercio dei vini a Barletta*. Barletta: Tipo-Litogr. Dellisanti e Giannone.

Ford, Alec G. 1969. Argentina y la Crisis de Baring de 1890. *Revista de Economía estadística* 13: 133–167.

Fortis, Marco, and Andrea Sartori. 2016. Food and Wine: Quality, Tradition and Innovation. In *The Pillars of the Italian Economy. Manufacturing, Food and Wine*, ed. Marco Fortis, 283–318. New York: Springer.

Fournier, Vincent. 2021. Intermediation and the Making of Cirò Marina Wine Markets in Calabria (Italy). In *The World of Wine, Networks and Scales: Intermediation in the Production, Distribution and Consumption of Wine*, ed. Stéphanie Lachaud, Corinne Marache, Julie McIntyre, and Mikael Pierre, 148–162. Bruxelles: Peter Lang.

Friz, Giuliano. 1980. *Consumi, tenore di vita e prezzi a Roma dal 1770 al 1900*. Roma: Edindustria.

Gabrielli, Patrizia. 2011. *Anni di novità e di grandi cose. Il boom economico tra tradizione e cambiamento*. Bologna: il Mulino.

Gallingani, Daniela. 1998. Gusti e disgusti: la tavola e il cibo nella letteratura e nel cinema. In *Storia d'Italia. Annali 13, L'alimentazione*, ed. Alberto Capatti, Alberto De Bernardi, and Angelo Varni, 929–932. Torino: Einaudi.

Garrabou Segura, Ramon. 2005. Conflict and Environmental Tension in the Adoption of Technological Innovation in the Agrarian Sector. In *Land, Shops and Kitchens. Technology and the Food Chain in Twentieth-Century Europe*, ed. Carmen Sarasúa, Peter Scholliers, and Leen Van Molle, 30–41. Turnhout: Brepols.

Girola, Carlos De Alberti. 1906. *L'Argentina Agricola: la sua capacità produttrice, le sue principali produzioni e le industrie derivate, importanza dell'allevamento del bestiame, il suo commercio, attrattive e vantaggi che presenta quale paese di immigrazione*. Milano: tip. Agraria.

González Esteban, Angel Luis, Vicente Pinilla, and Raúl Serrano. 2016. International Agricultural Markets after the War, 1945–1960. In *Agriculture in Capitalist Europe, 1945–1960: From Food Shortages to Food Surpluses*, ed. Carin Martiin, Juan Pan-Montojo, and Paul Brassley, 64–84. London: Routledge.

Grando, Stefano, and Gianluca Volpi. 2014. Backwardness, Modernization, Propaganda. Agrarian Policies and Rural Representations in the Italian Fascist Regime. In *Agriculture in the Age of Fascism. Authoritarian Technocracy and Rural Modernization, 1922–1945*, ed. Lourenzo Fernández-Prieto, Juan Pan-Montojo, and Miguel Cabo, 43–85. Turnhout: Brepols.

Hames, Gina. 2012. *Alcohol in World. History*. London and New York: Routledge.

Hancock, David. 2009. *Oceans of Wine. Madeira and the Emergence of American Trade and Taste*. New Haven and London: Yale University Press.

Helstosky, Carol. 2018. Alimentazione. In *Storia d'Italia. Annali 27. I consumi*, ed. Stefano Cavazza and Emanuela Scarpellini, 126–151. Torino: Einaudi.

James, Harold, and Kevin H. O'Rourke. 2013. Italy and the First Age of Globalization, 1861–1940. In *Oxford Handbook of the Italian Economic History since Unification*, ed. Gianni Toniolo, 37–68. Oxford: Oxford University Press.

Jones, Geoffrey, and Emily Grandjean. 2017. *Creating the Market for Organic Wine: Sulfites, Certification, and Green Values*. Working Paper 18-048. Boston: Harvard Business School.

La cucina italiana, anno V, luglio, 1956, p. 478.

Le taverne alberghiere e la rivincita del vino, «Enotria», anno XXXVIII, 2, febbraio, 1939, pp. 69–74.

Lepratto, Livio. 2016. Tra metafora e identità. Il gusto del vino nel cinema italiano. *Itinera. Rivista di filosofia e di teoria delle arti* 12: 112–131.

Livat, Florine. 2019. Individual and Collective Reputations in the Wine Industry. In *The Palgrave Handbook of Wine Industry Economics*, ed. Adeline Alonso Ugaglia, Jean-Marie Cardebat, and Alessandro Corsi, 463–485. Cham: Palgrave Macmillan.

Losada, Leandro. 2009. *Historia de las elites en la Argentina: desde la conquista hasta el surgimiento del peronismo*. Buenos Aires: Editorial Sudamericana.

Luzzatto Fegiz, Pierpaolo. 1952. *Gli italiani e il vino. Inchiesta sulle abitudini e preferenze degli adulti eseguita per conto del Ministero dell'Agricoltura e delle Foreste*. Roma: Edizioni Doxa.

Macedo, Marta. 2011. Port Wine Landscape: Railroads, Phylloxera and Agricultural Science. *Agricultural History* 85: 157–173.

Mainardi, Giusi, and Pierstefano Berta, eds. 2018. *Il Grande Libro del Vermouth di Torino. Storia e attualità di un classico prodotto piemontese.* Canelli: OICCE.

Malpezzi, Frances M., and Willian M. Clements. 1999. Bread and Wine in Italian American Folk Culture. In *The Italian American Heritage. A Companion to Literature and Arts*, ed. Pellegrino d'Acierno, 109–114. New York and London: Garland Publishing.

Manetti, Giovanni, Paolo Bertetti, and Alessandro Prato. 2006. *Semiofood: comunicazione e cultura del cibo.* Torino: Centro scientifico editore.

Marescalchi, Arturo. 1925. *Le esportazioni del vino italiano.* Casale Monferrato: casa editrice fratelli Marescalchi.

Martinelli, Pablo. 2014. Von Thu¨nen south of the Alps: access to markets and interwar Italian agriculture. *European Review of Economic History* 18: 107–143.

Meardi, Francesco, ed. 1883. *Atti della Giunta per l'Inchiesta agraria e sulle condizioni della classe agricola. Relazione del commissario avv. Francesco Meardi (Provincie di Cuneo, Torino, Alessandria, Novara, Piacenza e circondari di Bobbio e Voghera)*, vol. VIII, t. I. Roma: Tip. Forzani e C.

Medina-Albaladejo, Francisco J., and Tito Menzani, Tito. 2017. Co-operative Wineries in Italy and Spain in the Second Half of the Twentieth Century: Success or Failure of the Co-operative Business Model?. *Enterprise & Society 18*(1): 32–71.

Mediobanca. 2020. *Indagine sul settore vinicolo 2020.* Milan: Area Studi Mediobanca.

Meloni, Giulia, and Johan Swinnen. 2018. Trade and Terroir. The Political Economy of the World's First Geographical Indications. *Food Policy* 81: 1–20.

Mitchener, Kris James, and Marc D. Weidenmier. 2008. The Baring Crisis and the Great Latin American Meltdown of the 1890s. *The Journal of Economic History* 68: 462–500.

Mocarelli, Luca. 2013. The Long Struggle for the Chianti Denomination. Quality versus Quantity. In *Typicality in History. Tradition, Innovation and Terroir. La typicité dans l'histoire. Tradition, innovation et terroir*, ed. Giovanni Ceccarelli, Alberto Grandi, and Stefano Magagnoli, 323–340. Brussels: Peter Lang.

Mondini, Salvatore. 1916. *Industria enologica. Produzione, commercio, regime doganale.* Roma: tip. Nazionale Bertero.

Mortara, Giorgio. 1925. *Prospettive economiche.* Città di Castello: Società tipografica Leonardo Da Vinci.

Navarini, Gianmarco. 2015. *I mondi del vino. Enografia dentro e fuori il bicchiere.* Bologna: Il Mulino.

Notizie e studi sull'agricoltura. 1892. *Notizie e studi sull'agricoltura.* Rome: tip. Nazionale di G. Bertero.

Nowak, Zachary. 2019. A Transnational Fiasco: Authenticity, Two Chiantis, and the Unimportance of Place. *Global Food History* 5: 5–24.

Oddy, Derek J., and Alain Drouard, eds. 2016. *The Food Industries of Europe in the Nineteenth and Twentieth Centuries.* New York and London: Routledge.

Ottavi, Ottavio. 1893. *Viticoltura teorica-pratica.* Casale Monferrato: tip. Carlo Cassone.

Ottavi, Edoardo, and Arturo Marescalchi. 1897. *Vademecum del commerciante di uve e di vini in Italia.* Casale Monferrato: Tip. Lit. Carlo Cassone.

Pastore, Riccardo. 2004. *Il marketing del vino e del territorio: istruzioni per l'uso.* Milano: FrancoAngeli.

Pedrocco, Giorgio. 1994. Un caso e un modello: viticoltura ed industria enologica. In *Studi sull'agricoltura italiana: società rurale e modernizzazione*, ed. Pier Paolo D'Attorre and Alberto De Bernardi, 315–342. Milano: Feltrinelli.

Pellechia, Thomas. 2015. *Over a Barrel. The Rise and Fall of New York's. Taylor Wine Company.* New York: State University of New York Press.

Petrini, Carlo, and Gigi Padovani. 2017. *Slow Food. Storia di un'utopia possibile.* Firenze and Milano: Giunti.

Pinilla, Vicente, and María-Isabel Ayuda. 2007. The International Wine Market, 1850–1938: An Opportunity for Export Growth in Southern Europe. In *Wine Society and Globalization. Multidisciplinary Perspectives on the Wine Industry*, ed. Gwyn Richard Campbell and Nathalie Guibert, 179–200. New York: Palgrave Macmillan.

Pinney, Thomas. 1989. *A History of Wine in America. From the Beginnings to Prohibition.* Berkeley and London: University of California Press.

Planas, Jordi. 2016. The emergence of winemaking cooperatives in Catalonia. *Business History* 58(2): 264–282.

Pomarici, Eugenio, Luigino Barisan, Vasco Boatto, and Luigi Galletto. 2019. The Prosecco Superiore DOCG Industry Structure: Current Status and Evolution over Time. In *The Palgrave Handbook of Wine Industry Economics*, ed. Adeline Alonso Ugaglia, Jean-Marie Cardebat, and Alessandro Corsi, 421–435. Cham: Palgrave Macmillan.

Quirino, Paolo. 1991. I consumi in Italia dall'Unità ad oggi. In *Storia dell'economia italiana. 3. L'età contemporanea: un paese nuovo*, ed. Ruggero Romano, 201–249. Torino: Giulio Einaudi.

Rayes, Agustina. 2018. The Other Side of Argentine Foreign Trade: Sources for the Study of Imports, 1880–1913. *Revista de Historia Económica / Journal of Iberian and Latin American Economic History* 36: 13–25.

Regalsky, Andrés M. 2006. Modernización, expansión y crisis: una aproximación a la historiografía de las finanzas, la moneda y el crédito entre 1860 y 1930. In

La Historia Económica Argentina en la Encrucijada. Balances y Perspect, ed. Jorge Gelman, 101–117. Buenos Aires: Lugar.

Resnick, Evelyne. 2020. Les nouveaux entrepreneurs du luxe: bâtir un groupe à partir d'un grand vin. In *La vigne et le vin. Transformation des filières et des acteurs*, ed. Théodore Georgopoulos et al., 213–227. Paris: Harmattan.

Ritrovato, Ezio. 2010. Sviluppo dei mercati vinicoli dopo il 1860. In *Storia regionale della vite e del vino. Le Puglie*, ed. Antonio Calò and Liana Bertoldi Lenoci, 531–550. Martina Franca: Edizioni Pugliesi.

Roccas, Massimo. 2003. Le esportazioni nell'economia italiana. In *Storia economica d'Italia. 3. Industrie, mercati, istituzioni. 2. I vincoli e le opportunità*, ed. Pierluigi Ciocca and Gianni Toniolo, 37–136. Roma-Bari: Laterza.

Rosano, Dick. 2000. *Wine Heritage: The Story of Italian-American Vintners.* San Francisco: The Wine Appreciation Guild.

Rossati, Guido. 1910. The Vintage in Italy and Trade with the United States in 1909. *Bonfort's Wine and Spirit Circular* 10 (Jan.): 73.

———. 1921. *Dopo un anno di proibizionismo negli Stati Uniti.* Enotria 5: 21.

Rumbarger, John J. 1989. *Profits, Power, and Prohibition: American Alcohol Reform and the Industrializing of America 1800–1930.* Albany: State University of New York Press.

Salvatici, Silvia. 1999. *Contadine dell'Italia fascista. Presenze, ruoli, immagini.* Torino: Rosenberg & Sellier.

Scarpellini, Emanuela. 2008. *L'Italia dei consumi. Dalla Belle Époque al nuovo millennio.* Roma-Bari: Laterza.

Segres, Yves, Jan Bieleman, and Erik Byust, eds. 2009. *Exploring the Food Chain: Food Production and Food Processing in Western Europe, 1850–1990.* Turnhout: Brepols.

Serpieri, Arrigo. 1930. *La bonifica integrale.* Roma: Istituto Poligrafico dello Stato.

Simpson, James. 2000. Cooperation and Cooperatives in Southern European Wine Production. *Advances in Agricultural Economic History* 1: 95–126.

———. 2011. *Creating Wine. The Emergence of a World Industry 1840–1914.* London and New York: Princeton University Press.

Soldati, Mario. 2011. *Vino al vino. Alla ricerca dei vini genuini.* Milano: Arnaldo Mondadori.

Somogyi, Stefano. 1960. Cento anni di bilanci familiari in Italia, 1857–1956. *Annali dell'Istituto Giangiacomo Feltrinelli* 3: 121–263.

———. 1975. L'alimentazione nell'Italia unita. In *Storia d'Italia. 5. I documenti*, ed. Corrado Vivanti and Ruggiero Romano, 841–890. Einaudi: Torino.

Sorcinelli, Paolo. 1999. *Gli italiani e il cibo. Dalla polenta ai cracker.* Milano: Bruno Mondadori.

Spahni, Pierre. 2000. *The International Wine Trade.* Cambridge: Elsevier.

Stein, Steve, and Ana María Mateu. 2018. Argentina. In *Wine Globalization: A New Comparative History*, ed. Kym Anderson and Vicente Pinilla, 291–322. Cambridge: Cambridge University Press.

Stranieri, Stefanella, and Paolo Tedeschi. 2019. The Role of Quality in Wine Production and Market: European Rules, CAP and New Technology. In *A History of Wine Production and Consumption in Europe, Vol. II, Markets and Trade in the Nineteenth and Twentieth Centuries*, ed. Silvia A. Conca, Stéphane Le Bras Messina, Paolo Tedeschi, and Manuel Vaquero Piñeiro, 255–274. Cham: Palgrave Macmillan.

Swinnen, Johan. 2018. *The Political Economy of Agricultural and Food Policies*. New York: Palgrave Macmillan.

Tedeschi, Paolo, and Manuel Vaquero Piñeiro. 2018. L'agricoltura italiana dagli anni '30 al nuovo millennio nelle ricerche dell'Istat: crescita produttiva e benessere. In *Trasformazioni e risposte economico-finanziarie della società italiana nel lungo periodo*, ed. Donatella Strangio, 141–168. Milano: FrancoAngeli.

Trentin, Pompeo. 1895. *Manuale del negoziante di vini italiani nell'Argentina*. Buenos Aires: Tip. Elzeviriana di P. Tonini.

———. 1903. *Piccola enografia italiana in rapporto colla esportazione*. Firenze: Barbera.

Unioncamere, ed. 2007. *Rapporto sul settore vitivinicolo 2007*. Roma: Retecamere.

Unione Italiana Vini, ed. 1925. *Annuario vinicolo d'Italia. Edizione 1924–25*. Milano: Tip. E. Gualdoni.

Unwin, Tim. 1993. *Storia del vino. Geografie, culture e miti dall'antichità ai giorni nostri*. Roma: Donzelli.

Vaquero Piñeiro, Manuel. 2019. The Italian Wine in Argentina (1861–1914): Market, State Intervention and Wineries. *Storia economica* 22: 329–358.

Vaquero Piñeiro, Manuel, and Luciano Maffi. 2018. The Diffusion of Italian Wine in the United States (1861–1914). *Rivar. Revista Iberoamericana de viticultura, agroindustria e ruralidad* 15: 176–196.

Vecchi, Giovanni. 2017. *Measuring Wellbeing. A History of Italian Living Standards*. New York: Oxford University Press.

Vence Conti, Agustina, and Eduardo Martín. 2016. Cuesta Precios y salarios durante la crisis de 1890 en Buenos Aires. *Revista de Historia Económica / Journal of Iberian and Latin American Economic History* 34: 267–294.

Verga, Andrea. 1873. *L'ubbriachezza in Milano. Memoria*. Milano: Tip. G. Bernardoni.

Willson, Perry R. 2002. *Peasant Women and Politics in Fascist Italy. The Massaie Rurali*. London and New York: Routledge.

Zamagni, Vera. 1998. L'evoluzione dei consumi fra tradizione e innovazione. In *Storia d'Italia. Annali 13. L'alimentazione*, ed. Alberto Capatti, Alberto De Bernardi, and Angelo Varni, 171–204. Torino: Einaudi.
Zaninelli, Sergio. 1974. *I consumi a Milano nell'Ottocento.* Roma: Edindustria.
Zanoni, Elizabeth. 2018. *Migrant Marketplaces. Food and Italians in North and South America Food Forging Identity in the Italian Diaspora.* Urbana: University of Illinois Press.

CHAPTER 5

Italy Tasting: Wine, Tourism, and Landscape

Since the end of the twentieth century, the multiplication of studies on global viticulture (Storchmann 2018) represented a part of the extensive literature exploring the centrality of food and wine in Western societies emerging from the Cold War. The socio-cultural phenomena linked to food progressively grew up and had a relevant impact on the media, fostering a plurality of reflections and positions as evidenced by the multiplication of studies and scientific journals (Albala 2013). The debate immediately became multidisciplinary including economics, environment, sustainability, tourism, agriculture, anthropology, and so on: analyses highlighted global movements contributing to the formation of a food culture erasing national barriers (Higgins 2018; Tiefenbacher and Townsend 2020) and the preservation of local identities, including food and beverages. In this context, the wine sector continued to experience a dialectic between global brands operating in international markets and consumers whose choices depended on the reputation or notoriety of the wines, strictly linked to their places of origin (Schamel 2009).

The debate, involving concepts as brands, tangible and intangible heritage, heritage or cultural assets as well as the future of viticulture and oenology, concerned (and continued to) the wine tourism and territorial development policies. Food and beverages in general, but the wine in particular, always belonged to travellers' expectations: also in the past, as in the "Grand Tour" of the modern age, travels and visits were planned with the aim to have the possibility of experiencing particular emotions, taking

M. Vaquero Piñeiro et al., *A History of Italian Wine*,
https://doi.org/10.1007/978-3-031-06097-7_5

part in tastings, visiting historic cellars or walking through vineyards. So, the current challenge for the terroirs is to make the new dimension of food culture and foodies an opportunity to realise an eco-compatible economic growth (Richards 2015).

Italian vocation for tourism took profit by the insertion of food and beverage in the attractions for tourists together to the landscape, art, and cultural cities. An intensive marketing and branding activity was organised to create places of experience and creativity. However, there were some problems such as standardisation, homogenisation, rising costs, and the negative impact on environmental balance and resource consumption. This explains why people called for ethical policies and behaviour to make tourism a sustainable and eco-compatible economic activity (Fennell 1999; Holden and Fennell 2013).

Concerning Italian wine, new identities were built by means of food and wine in a context where the luxury concepts and consumption practices changed. These changes involved wine which belonged to new luxury goods: wine was transformed by new ways of eating, new ideas on food and new fashions. Wine tourism grew with the restaurant guides and the request for new eating experiences (Asero and Patti 2009).

However, luxury food and beverages are not synonymous with excellence: other qualities are crucial for the new gastronomy, whose necessary requisites are changes in mentality and a renewal of taste, focused on the excellence of the food. So, wine consumption was a way to construct identities and to assert ideas, which brought about the evolution of mentalities.

5.1 Making Wine Culture in Italy

During the second half of the nineteenth century, the Italian state was fundamental in the building of an educational offer concerning wine studies. At the same time, some relevant winemakers such as Bettino Ricasoli started to modernise the Italian wine sector. Even if the full renewal of the Italian viticulture and the success of the Italian wines in the international markets arrived in the last decades of the twentieth century only, Italy represented the perfect country to give best wines a historical and cultural dimension that went far beyond their quality of the drink. During the "Belle Epoque", it was in fact difficult to escape the emotions of knowing that the vines of Vesuvius grew not far from the cities of Pompeii and Herculaneum, or that the wines of Florence were the same tasted by the great artists of the Renaissance. The ancient and the modern, myth and reality, ended up merging without distinction or line of discontinuity: unique sensations in the construction of unique atmospheres.

One of the pioneers of the cultural turning point in Italian viticulture was Marquis Leopoldo Incisa della Rocchetta (1782–1871). On his family real estate in Rocchetta Tanaro (Piedmont), the nobleman set up a huge vine nursery in 1852 to provide viticulturists with an easy way to obtain good-quality vine shoots. His nursery hosted over 300 varieties of Italian and foreign vines. In his catalogues, published between 1862 and 1869, he described the characteristics of his vast ampelographic heritage, providing information on the denomination, bunch characteristics, the most suitable soils and exposures, productivity, and the characteristics of the wine produced. Cabernet, Sauvignon, Gamay, Malbec, and among the Piedmontese grapes he recommended were Barbera, Dolcetto, Grignolino, and Nebbiolo. His descendants, almost a century later, kept alive his vocation for innovation: so, his great-grandson Mario transformed the Tuscan farm of Bolgheri into the cradle of one of the world's most prestigious Italian wines, Sassicaia (Mora 2016, 127–130).

Thanks to the creation of a network of oenology schools, young technicians and professionals in oenology emerged and modernised the Italian oenological culture. From the last decades of the nineteenth century, they accumulated a very rich repertoire of writings explaining that Italian wine required technical knowledge to be applied to the production of good wines and also the dissemination in the countryside of new practices to be applied by wineries. They were not able to move most winemakers away from the idea of producing a poor-quality product, without the requisites to be sold abroad and to gain a good reputation, but they learnt how to improve their wines and make them competitive in the international markets. It is worth recalling some of the main players who facilitated this change of course. Names that marked an oenological revolution that was taking its first and still uncertain steps in Italy at the beginning of the twentieth century.

The central point of this change was the school of Conegliano, a town in the Veneto, a region where an agricultural academy had been active since 1769 and where the Oenological Society of the Province of Treviso was created in 1867. Its technical director Antonio Carpené published a book a few years earlier entitled *Nozioni teorico pratiche di viticoltura e di vinificazione* (Theoretical and practical concepts of viticulture and winemaking). In 1876, in Conegliano, the first special oenology school in Italy for the practical teaching of oenology began to operate. At the same time, Antonio Carpenè, together with his partner, Angelo Malvolti, created the Carpenè-Malvolti company (De Leo and Favero 2018), one of the oldest Italian wine companies: it produced sparkling wines which were sold on foreign markets. As in the case of the Marquis Incisi della Rocchetta and other nobles (Antinori, Frescobaldi), the Carpanè-Malvolti demonstrated how the modernisation of Italian winemaking depended on a solid entrepreneurship improving quality and adapting tastes to the consumers' tastes and needs.

Another person of great oenological importance was Giovan Battista Cerletti. After a long stay outside of Italy to learn the modern productive methods used in the agricultural schools operating in France, Belgium, Austria, and Prussia, he was appointed by the Ministry of Agriculture to direct the experimental oenological station in Gattinara and to prepare the foundation of the related special oenology school. In the same years, Austria founded an agricultural school in Poreč (Istria) and in San Michele all'Adige (Trentino), a provincial agricultural institute directed by Edmondo Mach.

In 1887, Arturo Marescalchi began studying oenology at the Conegliano school. He was the main protagonist of the first generation of technicians who favoured the renewal of Italian oenology. After his studies, Marescalchi continued his vocational training in Bordeaux, where he attended the university lectures of Gayon and Millarder. In 1891, he created the Society of Italian Oenotechnicians in Pavia and moved to Casale Monferrato to publish with Edoardo Ottavi the oldest Italian newspaper dedicated to wine, the *Giornale vinicolo italiano*. This was the beginning of a fruitful activity as a scholar and writer of articles and books which strongly influenced the evolution of Italian wine culture and knowledge. In 1924, he edited *I vini tipici d'Italia*: more than 100 wines were reviewed, divided by region. For each one, technical data were provided, such as production areas, grape varieties used and organoleptic characteristics. Between 1931 and 1937, he published together with Giovanni Dalmasso *Storia della vite e del vino in Italia* (History of Vine and Wine in Italy): it was the first great popular synthesis of the history of Italian wine, from the ancient to the twentieth century. Wine in Italy was presented in terms of art, literature, and folklore.

Marescalchi was also responsible for the creation of the Associazione Enotecnici Italiani (now Assoenologi, that is the Italian association of oenologists) and of the journal *L'Enotecnico* (now called *L'Enologo*). He was a member of parliament for four terms (from 1919 to 1934), a senator (since 1 March 1934), and deputy minister for agriculture from 1929 to 1935. During his political career, he was involved in regulating the wine sector. In 1921, with other parliamentarians, he presented the first bill for the protection and promotion of Italian "typical wines": although the measure was not promulgated, Marescalchi's commitment indicated that in the 1920s there existed the idea of making a precise identification between the wine and its production areas. He also represented Italy at the meeting in Paris on 4–6 June 1923, at which it was decided to found the Office International de la Vigne et du Vin (O.I.V.).

The consolidation of a greater oenological awareness in Italy also depended on the great diffusion of scientific journals as well popular newspapers and magazines. Publishers specialised in the edition of agronomic manuals that were easy to understand, thanks to the use of drawings and graphs: this allowed readers to know new tools and machines and their use during the winemaking process. Illustrations proved to be an effective vehicle for transmitting technical knowledge, even in the case of readers who were illiterate or unaccustomed to dealing with long and complex written texts (Govoni 2011). In the wine publishing sector, the members of the Ottavi family had a great relevance: they specialised in research, scientific dissemination, publishing, and politics, always with particular reference to vines, wines, and the wine trade. The first Ottavi to play a major role in the development of the Italian wine sector was Giuseppe Antonio (1818–1885) (Marescalchi 1904). He was born in Corsica (France), where he attended school in Ajaccio: in his 20s, he moved to Paris to undertake agronomic studies, and he graduated from the Institut Agricole of Grignon. In 1843, his excellent training earned him the direction of one of the first Italian agricultural institutes: the one in Sandigliano. He remained in Piedmont for the rest of his life, except for a period as director of a departmental nursery in Corsica between 1859 and 1862. He studied the best solution for the agronomic and in particular ampelographic development. He highlighted the relevance of the winemakers' education, the dissemination of knowledge, and educational visits to the terroirs. In the second half of the nineteenth century, he worked at the agricultural institute of Casale Monferrato and, in 1855, he began to publish the magazine *Il Coltivatore* (literally "The Grower"), which represented a faithful reflection of Ottavi's teaching and publishing activities. He was able to build up an important network of relationships at a national level: he, in particular, organised numerous agricultural visits allowing him to better know the different Italian regions and terroirs and to strengthen the links between more innovation-oriented winegrowers. Between 1857 and 1875, the visits reached more than 200 locations all over the peninsula: the so-called zigzag excursions became a regular and popular column of *Il Coltivatore*. Among his ten sons, Ottavio (1849–1893) and Edoardo (1860–1917) stood out for their dedication to agronomic studies and their dissemination.

Ottavio Ottavi graduated from the Scuola Superiore di Agricoltura in Portici, and in 1875 he was the founder of the *Giornale Vinicolo Italiano* (literally "Italian Wine Journal"), which was exclusively dedicated to oenology and viticulture. Together with his brother Edoardo, he was also in charge of *Il Coltivatore* and in 1877 he became the responsible, in collaboration with Gaetano Cantoni, of the *Almanacco Agrario* (literally "Agrarian Almanac"), demonstrating his great gifts as a publicist and writer of agronomic information. He was also a skilled salesman of chemical fertilisers, agricultural machinery, seeds and American vines.

Edoardo Ottavi, graduated from the Scuola Superiore di Agricoltura in Milan, he went on to specialise at the École Nationale d'Agriculture in Montpellier. He was one of the fathers of Italian agricultural journalism, serving as director of "Il Coltivatore" and "Giornale vinicolo italiano". He promoted bold and modern reforms in cultivation methods and the dissemination of agronomic knowledge through the institution of the "Cattedre ambulanti di agricoltura" (literally "itinerant agricultural chairs") whose agronomists went to the rural villages to educate peasants (including viticulturists). He was a deputy of the Kingdom of Italy and the author of some highly successful works such as *Viticoltura teorico-pratica* (1893), *Principi della viticoltura* (1909), *Enologia* (1918), or *Malattie e alterazioni dei vini* (1927), that is books dedicated to the winegrowing, winemaking, diseases of vines and adulteration of wines. During the 1890s, he began to collaborate with Arturo Marescalchi, husband of his sister Carolina: they maintained the editorship of *Il coltivatore* and two relevant series of technical monographs and manuals, that is the *Biblioteca agraria* (literally "Agrarian Library") and the *Biblioteca minima Ottavi* (literally "Ottavi's minimal library"). In 1902, they edited the wine guides to Tuscany, Sicily, and Piedmont: they in fact understood the relevance of the nascent motoring movement, providing practical information for motorists (maps, services, places to stay and eat). These guides de facto represented the first innovative experiments in Italy in the field of the wine tourism. Other publications having a particular focus on oenology and viticulture were the volumes published by the Ulrico Hoepli publishing house in Milan (Decleva 2001), an enterprising bookseller-publisher who edited numerous highly successful manuals on Italian ampelography and vine diseases.

During the twentieth century, the diffusion of scientific journals and magazines concerning the wine sector continue favouring the dissemination of the ampelographic knowledge in the countryside and training of oenologists who are also experts in disseminating information. One of the most important was Paolo Desana: he was in fact elected to the Senate with the Christian Democrat party for the third republican legislature (1958–1963), and he was the first signatory and main supporter in the Senate of law 930/1963, that instituted the Designation of Origin of Italian wines (Benatti and Castelletti 2022). Paolo Desana also served as president of the "National Committee for the Protection of the Denominations of Origin of Wines" (1966–1989) and, with the passionate collaboration of the most important oenologists and winemakers, he created over 200 DOC/PDO Italian wines. As journalist, he contributed to the *Corriere Vinicolo* (literally "Wine Courier"), *Civiltà del bere* (literally "Civilization of Drinking"), and *Il Coltivatore* and *Giornale Vinicolo Italiano*, and was the author of articles concerning the problems of the Italian wine sector and the solutions he indicated (many of them were exactly his much-appreciated speeches in the Senate).

All these protagonists of the evolution of the Italian wine sector showed both a firm belief in the economic potential of Italian wines and a great ability to disseminate their wide knowledge concerning vines and wines: other relevant oenologists gave, instead, more attention to improving the image of Italian wine in the world in order to increase sales in international markets.

5.2 Brand and Image of Italian Wine

In the construction of the political and economic identity of post-unification Italy, the wine had a prominent role (Guy 2003). Italy had no lack of raw materials and, compared to France, had a greater ampelographic diversity. However, commercially and reputation-wise, with a few exceptions, the picture was bleak. Sales outside Italy were low and many Italian wines were completely unknown to consumers living in other European countries: this depended on low quality and backwardness of the production systems used. In the international wine market, Italy was a simple exporting country for bulk and blended wines (Pinilla and Ayuda 2007). So, a radical change was needed to make sales of quality wine in the foreign markets a significant item in the balance of trade. In this context advertising proved to be a key tool (Falabrino 2007) in opening up new markets and creating a more developed image of the country. In 1872, one of the most famous brands of Italian beverages in the world was registered, the Martini of Turin: ten years later it began to use its most famous label, which progressively became an authentic icon of Italian products.

In Tuscany, Baron Bettino Ricasoli was responsible for the effort to obtain the "perfect Italian wine" during the nineteenth century: he transformed his castle of Brolio into an oenological laboratory where Chianti became the "true type of red table wine" suited to the taste of consumers and the demands of the long-distance wine trade. After the results achieved by Ricasoli, the opinion spread in the country that it was possible to obtain accredited table wines, putting an end to the custom of selling off poor-quality wines.

In Piedmont, too, wine producers became protagonists of such an operation: at the end of the 1870s, large wineries such as Martini or Gancia increased their dimension, separating the agricultural part from the industrial part and requiring farmers to cultivate particular grape varieties. In 1878, the wine company founded by Count Emanuele Alberto, son of King Vittorio Emanuele, was the first one rationally set up to produce and enhance Barolo wine, even using bottles with the name of the grape variety written on the label (Cantino et al. 2016). In 1892, the E. di Mirafiore company registered its trademark for the production and marketing of high-quality wines as Barolo, Barbera, and Freisa. It was the first step of a revolution with concrete repercussions in the field of marketing and the construction of a family business brand identity (Conz and Magnani 2019).

At the end of the nineteenth century, the old cellars, often unsuitable for the production of high-quality wines, began to be replaced by production facilities organised as if they were modern establishments. Building complexes were also depicted in posters, playbills, and postcards as a symbol of the modernity achieved. The modern plants became a real vehicle of commercial communication useful for conveying both the achievements of the Italian economy and the spirit of enterprise of a modern, active wine business. As postcards, letterheads, posters, newspapers, almanacs, and tourist guides attested, the architectural choices also reflected the image of the renewed Italian wine sector.

The creation of a new generation of wine cellars, in many cases built with the help of engineers, was an important innovation which visually marked the transition from the "artisan" winemaking to the "industrial" one. The new rational arrangement of the rooms, in addition to elegance and order, allowed to improve the wine quality and represented the modernisation of the productive system. The advertising, as in the case of the Dietz company in Naples, highlighted the new different way of doing business in the wine sector: the message for consumers was clear, that is best wine companies used modern winemaking to realise high-quality wines.

Best Italian winemakers had to learn how to produce and sell on a global scale while respecting standards of quality and emotional response: it was imperative that their products had precise recognisable traits, attracting the interest of customers, but also reassuring them about quality and hygiene. From this point of view, the story of advertising and in general of the image to be assigned to wine and other liqueur drinks is an excellent case study of the transformations experienced by the Italian agri-food industry.

During the "Belle Époque", new Italian wines had to be equipped with elegant labels that conveyed the aesthetic qualities of their terroirs of origin: landscapes and historical buildings were used for labels and advertising posters, without neglecting other elements of great visual impact such as works of art, industrial plants, and local folklore recalls (Grazioli 2001). Besides the images concerning technological innovations, accurate portraits of pretty peasant women caught at the moment of completing the harvest illustrated that Italian wine arrived from a land full of human and natural beauty. Consumers in the international markets had to think that Italian wines were the result both of the skilful work of skilled winemakers and the unrepeatable goodness of the natural environment.

The birth of a new generation of wine entrepreneurs in Italy, at the end of the nineteenth century, represented an innovation of considerable importance. France had a consolidated advantage, with regard to wine production, and in terms of sales: French winemakers demonstrate a greater ability in adapting the characteristics of their products to different target markets. In Italy, during the 1880s, a label on a bottle was described as a "card" and treated as an object of curiosity, an example of the considerable cultural distance that existed between the two countries. Operators unsuited to this kind of business were predominant. They lacked the essential requirements needed to successfully promote a product as wine on the international stage. At the beginning of the twentieth century, the wine trade in Italy was dominated by local intermediaries, while buyers normally tasted the wine before agreeing a price. Foreign markets required other commercial practices, less tied to personal relationships and more based on product reputation.

Few Italian wine companies were concerned exclusively with the distribution of wine. As shown by newspaper advertisements in major foreign cities (Buenos Aires, New York, London), best winemakers began to work with agencies with specialised staff that kept their deposits in proper conditions. Cella Bros in New York, Charles Ciceri in Montréal, Fratelli Romeo in New York, and Vincent Bosco in New York were Italo-American companies founded in the late 1890s and early twentieth century managing a commercial area that was still being defined. The US market was a test case, as the growing production of ordinary wine in California indicated that Italy could gain an advantage by concentrating on the export of high-quality typical wines, the only product able to compensate for the high cost of transport and customs duties (Vaquero Piñeiro and Maffi 2018). As the consumption of alcoholic beverages was closely linked to the disposable income, it was possible to conquer the US market by concentrating on quantity or on quality: however, it was necessary to adapt Italian wine companies to new marketing and advertising techniques capable of orientating the consumption of the sectors of society with greater spending power.

This was the case, for example, of Pontin's Restaurant, a well-known New York establishment advertising Italian cuisine served in an authentic Italian style garden in the pages of the Italian chamber of commerce magazine in New York, just a few meters from the courts and trading companies. Along with a wide range of French and German wines, the wine list also included a selection of Italian wines, above all Chianti, but also Muscat, Nebbiolo, Lacryma Christi, and other wines from the south of Italy. For the observers sent to the United States by the Italian government, it was not enough to merely learn the entrepreneurial techniques implemented by Californian wineries. It was fundamental to guarantee a place for Italian wines on the menu of luxury restaurants: the success with their wealthy clientele represented the key for the penetration of Italian wine in the US market.

It was also necessary to master the nascent techniques of advertising, as for wine, a product rich in social and cultural values: to increase sales, it was not enough to be a good-quality and well-priced wine. Besides, a recognisable name, in association with the scenic and artistic characteristics of the Italian regions and terroirs, was considered an excellent promotional vehicle for Italian wine (and in general for all Italian foods and beverages). Enological *réclame* became one of the topics of debate among operators in the wine sector: the progressive increase in exports of bottled wine needed a differentiation of products with a label, featuring the name of the wine and a captivating brand. As was the case with other products, such as pasta, oil, mineral water, and tinned tomatoes, Italian wine stimulated changes to adapt to new requirements imposed by the internationalisation of the language of advertising. Italian products remained far from American standards of brand recognition (Fasce 2012), but in particular for the wine, they started to deal with the changing tastes of customers living in distant markets. To meet and satisfy the lifestyles of the emerging classes of the Belle Époque, and to overcome the traditional view of wine as a popular drink, advertising was entrusted with the creation of a stimulating image for quality wine and the attribution to wine of a distinguished position in restaurants, official lunches and, not least, in the daily life of the family. While the French wine had access to a large and proven repertoire of evocative *chateaux*, a testing ground was created for Italian graphic artists, combining realistic and allegorical elements.

The Italian bottles had to be elegant, with meticulous attention to detail in the presentation and labels capable of reflecting the best Italian artistic tradition, drawing inspiration from Italy's many natural wonders. The illustrators called upon to design the image of such a symbolic product had a wide range of themes at their disposal, from which to draw inspiration: the coats of arms of royal houses, floral decorations, works of art, ancient buildings, mythological scenes, and most of all beautiful scenery, such as Vesuvius, the Gulf of Naples, and Mount Etna, all of which were easily identifiable. The landscape became both the instrument and object of communication, helping to create the typical image of Italian wine: to remind Italian tradition, the advertising image evoked the origin of a product that did not come "from the dark bowels of the ground, nor from the smoky and grey neighbourhood workers in the big cities, but rather from free nature, from the plains and hills" (Rocchi and Gabbai 2013). So, wine advertising was transformed into a story in which allegorical images presented Italian wines as elements of an almost mythical, dreamed world, transporting the consumer to a sort of commercial Eden.

During the 1920s and 1930s, many of the trends previously developed slowed down due to the autarkic policies imposed by Fascism, also with regard to food and drink consumption. In addition, trade relations with foreign countries became less intense and the role of advertising lost its importance. The Italian winemakers had to address themselves largely to the domestic market: so, they adopted a language much less rich in graphic suggestions and resources, at most recovering elements clearly identifiable with the most evocative pages of Italian history. Advertising made extensive use of coats of arms, castles and graphics to establish an ideal link with the most evocative pages of the Italian medieval and Renaissance. The fascist regime showed great interest in forms of visual culture: they in fact helped to link Mussolini's Italy with the most glorious pages of the Italian ancient history (Lasansky 2004; Lazzaro and Crum 2005). In exalting a drink that was a typical expression of Mediterranean culture (Montanari 1995), Italian wineries contributed to the construction of a collective imagination where consumers thought to the wine of imperial Rome or the courts of the medieval seigniories: the paradox, existing in all advertising referring to the past, was that in reality those ancient wines were in reality absolutely not drinkable for the modern consumers' taste.

Fascist regime also imposed the "grape festival" with the declared aim of encouraging the consumption of table grapes: between the months of September and October, in all Italian towns, the population was invited to take part in parades with floats and patriotic celebrations in homage to the fruits of the Italic land (Cavazza 1997).

Some novelties also concerned the emerging link between wine and tourism. The practice of sport and propaganda on holidays in the mountains, at the seaside or in seaside resorts in contact with the environment and local folk traditions (Berrino 2011), favoured the creation of wine routes in Italy following the German example (the first didactic wine road, Weinlehrpfad was in fact created in Germany at the beginning of the 1930s). *Le vie d'Italia* (literally "Italy's ways") edited by the Touring Club and other magazines aimed at the high society having the possibility to make expensive holidays and leisure trips recommended that large hotels or alpine refuges would allow to taste the best national and international wines. This concerned few people, but the idea to create a marriage between wine, tourism, and territories was definitively emerged.

The construction of a more evocative image of wine was put aside with the Second World War and the years of economic recovery: however, the existing rich heritage did not disappear completely and remained ready to re-emerge and gain visibility during the 1950s and 1960s. The "economic miracle" and the related increasing incomes of Italian families determined the definitive entry of Italian society into mass consumption: even if the advertising, previously reserved for a small minority of tourists and high-status consumers, progressively changed adapting itself to the new mass market, some basic elements remained constant in the construction of a long-lasting storytelling around the relevance of wine for the Italian territory and culture.

This process de facto started when also the Marshall Plan, strongly reducing aids in cereals and raw materials and increasing the sending of new advanced technological tools, indicated that the emergency was finishing (Fauri and Tedeschi 2011). In 1950, taking advantage of the celebration of the Holy Year in Rome and in anticipation of the arrival of thousands of foreign tourists, luxury hotels and restaurants employed qualified staff who knew how to serve wine correctly, providing precise information on the qualities and areas of origin. The *Corriere Vinicolo*, a newspaper of the "Unione Italiana Vini" (literally "Italian Wine Union"), invited winemakers to meet customers, creating convivial occasions to increase sales. Winemakers had to learn the marketing and communication techniques imposed by tourism: they had to organise trips to wine-growing areas and to wine cellars where receptions with tastings were prepared. They also had to collect the addresses of tourists to send them complimentary packages and started to retain new customers. In this climate of relaunching the Italian tourism sector (Scarpellini 2011), the Italian Centre for Tourist Development published the multilingual guidebook *Turismo Gastronomico* (literally "Gastronomic Tourism") which was linked to the Gastronomic Guide to Italy published in 1931 by the Italian Touring Club.

Even if these recommendations were not understood in a context still dominated by a rural vision of wine production, they clearly indicated novelties that were slowly appearing, thanks to best winemakers (Gabrielli 2011). For example, the Barolo labels included for the first time recommendations to be followed in order to savour a wine of excellent quality: "Barolo, the gem of Italian production, is the most aristocratic and the finest roast wine, universally known. It should be poured gently, taking care to uncork the bottle a few hours before serving at a temperature of 18 degrees". A new relationship was established with the consumption of wine, which ceased to be a necessity and became a moment of pleasure.

Best winemakers in particular followed the example arriving by the Californian wineries: as it normally happened for many US products, in the 1950s, the emulation and the desire to become rich as US citizens pushed Italian entrepreneurs to observe US strategies. Concerning the wine sector, the example of Californian wineries organising visit to their wine was relevant and fuelled the existing the myth of the educated and "mechanised" Californian winemaker, a myth diffused in all Western European countryside and justified by the experience they had acquired since the 1930s. In order to overcome the consequences imposed by the restrictive prohibitionist laws of the 1920s, Californian winemakers of Napa Valley had progressively introduced the fashion of opening their cellars to wealthy customers from San Francisco and other cities of the US Pacific coast (Gmelch and Gmelch 2011; Walker 2005). They also had created lively and colourful advertising posters praising the beauty of the Californian lands bathed in sunshine and covered with lush rows of vines. Their message was clear, as bottles of wine appeared in the foreground, emphasising the goodness of the local landscape and productions. So, in the 1950s, the advertising messages about Californian wine showed Italian winemakers how to illustrate the close relationship between the qualities of the terroirs and its main product.

The new lifestyles guaranteed by the increasing Italian earnings of "the economic miracle" imposed different behaviours in the Italian society, in particular in the cities (Cavazza and Scarpellini 2010; Gabrielli 2011). As an expression of modern times resulting from rising incomes, advertisements indicated that the wealthy classes of the cities needed to know the characteristics of wines in order to pair them with the right dishes, especially when taking part in some important dinner or reception. As the magazines and guides suggested, Italian wines had to be identified with a name and a specific area of origin because their consumption required special skills. Customers began to visit the terroirs and then consumed the wine at home or in restaurants (Image 5.1).

New wine routes were another valuable vehicle for promotion. The first examples of wine roads arrived in Friuli, in the mid-1970s (*Strada del Vino nel Collio* and "Strada del Merlot") (Costantino and Artista 2003). In 1993, new roads started throughout Italy: although the law formalising them was promulgated some years later, "wine routes" immediately became an excellent tool to promote tourism not only in relation to wine but also to the territory producing it (Antonioli Corigliano 1999). Italian legislator enacted a specific national law (27 July 1999, n. 268) for the wine routes whose aims could be extended to other typical products such as oil and cheese: in the Article 1, they were defined as "(…) paths marked and advertised with appropriate signs, along which gather natural, cultural and environmental values, vineyards and cellars of single or associated farms open to the public; they represent an instrument through which the

Image 5.1 The antique façade of a wine and olive oil store in Rome, Italy. (Source: © foto di Gabriella Clare Marino. https://unsplash.com/photos/IRqjAsYU0B4)

territories and the relative productions can be disseminated, commercialised and enjoyed in form of tourist offer". The main goal of the law was to promote territories with a wine vocation, with particular reference to those with high-quality productions. It highlighted that a route was an integrated system of both territorial and tourism offers organised along a route characterised by places of historical, artistic and environmental interest, flanked at the same time by a series of structures of reception, promotion and marketing of local products. New law provided the necessary prerequisites in order to increase the competitiveness of the territorial systems and to contribute to the formation of the roads themselves, by aiming at creating effective networks characterised by a bottom-up approach: this last required active partnership as well as planning skills on the part of the public and private operators belonging to both the chain of typical products and the tourism sector.

Also in the following decades, there was an increasing development of territorial brands (Higgins 2018): the identification of corporate brands with specific contexts with a strong tourist vocation was strengthened. In particular, industrial and financial groups invest in wineries to associate their names with particularly famous Tuscan wines. A phenomenon defined as "one of the most widespread and complex commodities produced globally" (Inglis and Almila 2019) that stimulated an abundant scientific literature as a result of the intertwining of social, cultural, and economic factors. Increased ecological awareness, movements in favour of a return to agriculture, and the growing importance of wine tourism, all contributed to bringing attention back to territories, as well as to individual entrepreneurial realities. The phenomena linked to wine tourism and the search for authentic wine landscapes began towards the end of the twentieth century and then came UNESCO recognition (Sigala and Robinson 2019).

Food and beverages and the culture of the territory progressively became a harmonious conjunction between the tourism and consumers' expectations and interest both in the gastronomic product itself and in all the aspects used by advertising to visually narrate the existing union between landscape, history and culture (Tiefenbacher 2013). The promotion of local wine highlighting the territorial brand also contributed to the economic development of the rural territory and socio-economic well-being through the growth of gastronomic tourism: this last was linked to the landscape, local cuisine and the culture from which it originated (Charters and Spielmann 2014). All this has led, in the Italian case, to the inclusion of some terroirs in the list of UNESCO World Heritage Sites: this sometimes contributed to favouring an immobile vision creating some difficulties in terms of introduction of innovations and quality improvements. Besides, it was also evident that UNESCO heritage sites, including those in the wine sector, fuelled debates about the narrative used to justify the resources spent on enhancement efforts to obtain this international recognition (Balzani 2015). However, the advantages of the inclusion were superior to negative effects, in particular when local wine companies and public administrations created business museums and eco-museums strongly identified with the terroir and the winegrowing and winemaking (e.g. the case of Chianti). Thanks to this, the territorial brand has been consolidated as a vehicle for economic growth. (Magliacani 2015, 68–71).

Italian terroirs becoming UNESCO patrimony for vine landscapes were in 2014 the Piedmontese Barolo area in the Langhe, Roero, and Monferrato, which was followed in 2019 by the Prosecco hills of Conegliano and Valdobbiadene (Visentin and Vallerani 2018). To these two areas with a specific wine vocation, it is possible to add the UNESCO territories for the landscape in general where the viticulture offered a substantial contribution: Castel del Monte (Apulia), the Amalfi Coast (Campania), and Val d'Orcia (Tuscany). In this way, the Italian wine sector participated in a worldwide phenomenon based on the principle of welding the relationship between human action (winemakers and consumers) and the environment (Sigala 2019). In this context, the wine tourism progressively became a real economic, social, and cultural heritage: the territorial brand sent a message of quality assurance that extended from wine to the entire cultural ecosystem near the terroir. In 1963, with the law on designations of origin, Italian wines of the different terroirs acquired a definitive identity. In the new millennium wine, as a symbolic product, became one of the determining factors in the construction of the nature of the geographical areas of production (Crescenzi et al. 2021) (Fig. 5.1). This explains why Italian consumers identified (and continue to) wines with the region/terroir of origin more than with company brands (Corsi et al. 2019): places were more important than winemakers.

The creation of specific advertising for Italian wine had to include all aspects of the wine chain supply, from the vine cultivation to the wine consumption through the winemaking, the marketing and the mass media (Mazza 2010, 63). Billboards, posters, advertisements and, in more recent

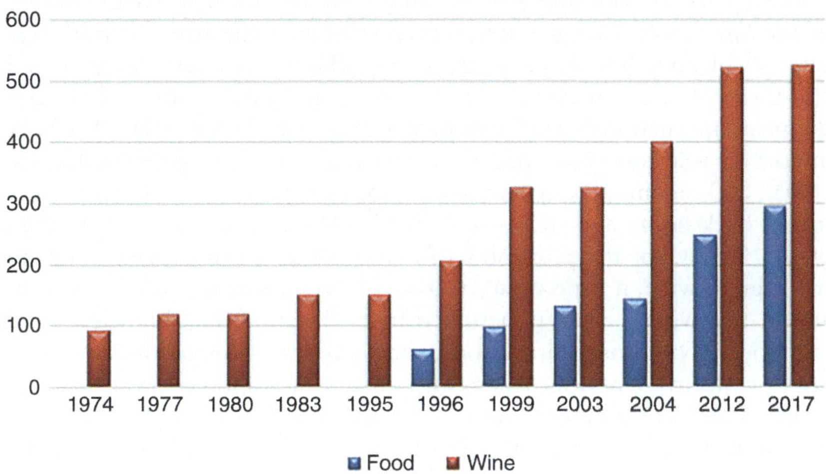

Fig. 5.1 Trend of Italian geographical indications (wine and food). (Source: Our elaboration from www.qualigeo.eu/statistiche-eu-dop-igp-stg/)

phases, social media, were used by an entrepreneurship obliged to constantly renew the narrative surrounding the wine and its strong links with the consumers belonging to all social classes. All changes, even if they were necessary, were difficult to adapt to the most traditional of Italian beverages, and they met many socio-cultural resistances: so, the advertising helped to promote the renewal. In this context, the use of landscape and the natural environment reassured traditional consumers while involving new ones: so the consolidation of a modern wine sector was presented with the images of wine landscapes becoming vectors of a new territorial identity and of the construction of an adequate tourist offer (Harvey et al. 2014).

Talking about consumption and demand for the wine, a product so full of subjectivity and social values implied (and continue to) the need to look beyond the simple statistical trend of the productions and sales. Numbers do not always tell the whole story. Italian sector wine faced the "long transition" of Italian society from food scarcity to abundance: best wine companies used image, advertising and mass communication techniques, and territorial marketing strategies projected onto competitive global scenarios until arriving at the current concept of brand and capacity building for (Da Silva Lopes 2007). The last aspect of the wine chain supply, from producer to consumer, assumed an important role: wineries and vineyards, but also places of consumption, contributed to the formation of the beauty of the landscape. The geometric arrangement of the rows of vines was carefully planned, the shapes of the vines and buildings were designed and also their colours were chosen. In the marketing and promotion strategies of main international wine companies, the wine cellars played a very important role too. They had to attract the interest of globalised, individual, and organised tourists, looking for new emotions and experiences to tell back home and, in current times, on social networks (Kernohan 2014). Today, for the post-modern consumer, it is no longer enough to buy a good wine, it is necessary to know how to taste it and tell about it, almost following a ritual punctuated by explanations and knowledge: all with the aim of transmitting a social fact linked to food and drink consisting of collective emotions (memories, images, testimonies to be shared). From alcohol consumption out of necessity to luxury oenophilia as a lifestyle, Italian wine was a protagonist of this incredible transformation and became an iconic representation of Italian history, habits, and style (Resnick 2008).

During the twentieth century, the debate on typical wines divided "techno-oenologists" and "purists": for the first ones it was important to produce "constant" wines having the same characteristics even if they were realised with the fruits of different grapes harvest: for the second the standardisation of flavours was fine for so-called plebeian or bourgeois wines, not for the high-quality ones. There existed two incompatible positions: "techno-oenologists" thought that the "constant" taste and flavour had to be typical attributes of high-quality wines—many wineries could increase the quality of their products and enlarge their markets. For the "purists", wineries producing the "aristocratic" wines, today "luxury wine brands" (Yeung and Thach 2019), had to realise unrepeatable wines, giving theme, depending on the vintages and barrels, precise characteristics of uniqueness and obviously varying year after year: they wanted to always guarantee a high quality, but not necessarily the same. Standardisation of flavour and taste in fact risked to reduce the value of the greatest wines and confused them with other ones having no prestige. In the early new millennium, Italian wines started to occupy a solid position in the world rankings of the most prestigious wines: however, "techno-oenologists" and "purists" continued to debate, but both of them produce high-quality wines. Some Italian wines assumed the value of an asset of refuge for major world investors: so, they were bought as an investment and not for drinking (Navarrini 2015). In 2010, there were 8 Italian wines in the prestigious LIVEX ranking of the world's most important wines: before the Pandemic Covid-19 they became 17 and 4 were in the top ten: Gaja (position 3), Sassicaia (4), Ornellaia (6), and Masseto (9). As the ranking is drawn up according to a set of parameters including trading volumes, price trends, the number of labels handled per brand, and so forth, this result clearly indicates the success of the Italian wine sector in the world. For Italian wines, 2020 was in fact a great year of progress: prices grew on average by 6 per cent, more than for all other countries (whose average was 3.2 per cent), and the number of wines traded was around 25 per label on average, still below the French average but in constant progress. The price in absolute terms remained on average one of the lowest (around £ 1700 per 9 litre case): however, an Italian wine that cost 100 euros in 2009 arrived at 235 euros in 2020, equalling the performance of Bordeaux (€236) and overtaking both Champagne and Rhone (around €215).

5.3 Exploring Wine Tourism and Landmark

Long before concepts such as "short supply chain", "eco-sustainable tourism", or "slow tourism" (Clancy 2018) were normally used in the new millennium, a debate based on the concept of land as a "productive factor" began during the 1980s. Economics and ecology met: the interests of production could coexist with the problems of environmental conservation. To achieve this goal, the path of planning was followed. Law 1497 issued in 1939 already spoke of protected assets in defence of the "scenic beauties", as they were called at the time. In 1970, the statutes of the Italian regions incorporated the principle that regional authorities were required to "promote a balanced territorial organisation aimed at creating an environment capable of fully satisfying the needs of the human person. It provides for the protection of the soil and the landscape, the protection of natural resources, the ecological environment and the historical, artistic and archival heritage". So, during the 1970s new laws on nature parks, mountain areas, the protection of the natural environment and the agro-forestry heritage were promulgated. Besides, in the 1980s, urban-environmental development plans and projects were issued: they addressed land use according to its specific potential and peculiarities. In particular, the criterion of maintaining the use of the most valuable agricultural land was introduced. New legislative interventions outlined "in the various areas subject to protection restrictions, not only prohibitions to do, but also positive vocational and planning opportunities": these latter obviously had to be supported by appropriate incentives, contributions and tax breaks. Rural tourism or agri-tourism was in particular regulated, in 1985, by "Legge quadro sull'agricoltura" (literally "Framework Law on Agriculture"), according to which accommodation was to be developed on farms: the aim was to emphasise the agricultural side in the tourism context.

This means that the attention to sustainable development was present in Italian laws before it became so important in all economic and ecologic debates: in particular the concept that tourism did not necessarily mean unbridled consumerism was strengthened. It could be less invasive from the point of view of numbers and also more attentive to respect and protection of the environment: it was the discovery of non-mass tourism (Battilani 2010). Agricultural activities were considered relevant resources for the regional development and new forms of rural tourism or ecotourism (Fennell and Weaver 2008; Fennell 2013) became tools for the development of the countryside, as they increased the income of peasant families. In the early 1990s, this trend was increased by the McSharry reforms concerning the Common Agricultural Policy (CAP): the EU Regulation 2078/92 in fact dealt with the relationship between agriculture and the environment also favouring the agri-tourism in the European Union by fiscal facilities. Funding was granted to farmer-entrepreneurs who committed to environmentally friendly farming practices, including the creation of farms dedicated to the agri-tourism.

In 2000, the European Landscape Convention was approved, and in 2004, the Italian Code for Cultural Heritage and Landscape was adopted. Although there was no definition of "landscape", new European, national, and regional laws consolidated in public opinion the value of landscape as a resource to be handed down intact. Landscape had to contribute to the social well-being of the population as a factor enhancing the quality of life and strengthening cultural identity, as well as increasing the activity and economic competitiveness of the areas concerned.

The official recognition of wine tourism only arrived in 2019: wine tourism was defined as "all the activities of knowledge of wine carried out in the place of production, visits to places of cultivation, production or display of tools useful for the cultivation of the vine, tasting and marketing of the company's wine production, also in combination with food, initiatives of an educational and recreational nature within the wineries". In the meantime, Italy was preparing to host the Global Conference on Wine Tourism of the World Tourism Organization, the United Nations agency responsible for promoting sustainable and responsible tourism, scheduled for 2022.

These changes concerning legislation confirmed the increasing relevance of the wine tourism (or oeno-tourism) for the Italian economy, in particular for the countryside (Boatto and Bianchin 2011): it was evident that the growth of the number of tourists attracted by the countryside and in particular by the wine tourism could guarantee new earnings, while the revenues concerning crops granted by the CAP decreased (Tedeschi 2011). The wine tourism was added to the traditional concept of summer holidays associated with the sun and beach binomial (Battilani 2001) allowing European families an alternative to spend their free time. So, wine tourism became an example of the possible interconnection between environment, culture, and rural society (Cavaco Carminda 1995; Galletto and Bianchin 2011). The landscapes with vines and the building where winemakers produced their products in fact contributed more than other agricultural sector to show a beautiful tourist image of the countryside: so they represented a concrete factor of growth for the rural economy, that is the wine and wine tourism were the main pillars of the agri-food system in Europe. Also from the perspective of "experiential tourism", wine immediately appeared as one of the landscape factors that contributed to the transmission of values (Carlsen 2004). In the new debate considering the advantages and criticalities (economic, environmental, social) of the "marriage" between the wine sector and the tourism (Hall et al. 2000; Lacoste and Zamora 2007), a new definition of wine tourism was also indicated: "visitation to vineyards, wineries wine festivals and wine shows for which grape wine tasting and/or experiencing the attributer of a grape wine region are the prime motivating factors for visitors" (Carlsen and Charters 2006, 1).

Not only in European countries with a traditional winemaking, but also in many other emerging countries, wine progressively became the centre of a wide range of activities and interests, from new marketing and sales techniques to consumer habits or the creation of surprising buildings. So, in the last years 53 per cent of world travellers claimed to consider themselves food and wine tourists. According to the Italian food and wine tourism report for 2019, 56 per cent of Italian tourists claimed to have visited a winery, while 44 per cent claimed to have attended wine-themed events or festivals. The wine sector favoured the emergence of a modern agribusiness unifying the tourism with agricultural activities and productive systems.

Differences emerged between the trajectories being pursued by countries having a young winemaking history (Anderson and Pinilla 2018) and the European countries whose wine sector was the object of a relevant agricultural reconversion in the new millennium. In these latter and in particular in Italy, the terroirs and wine routes were modified to include many other tourist attractions, from typical food products, to festivals, to the landscape and architectural attractions of the localities involved. The images originally intended to be a tool for promoting the wine sector started to help the advertising needs of all products of the countryside.

It was during the last decades of the twentieth century that the wine sector found in the new food and wine fashion a secure anchorage point (Richards 2002): the first wine shops with an extensive list of regional wines were born, thanks to the "sponsorisation" of new associations. In 1987, the *Associazione Nazionale Città del Vino* (literally "National Association of Wine Cities") was founded: its aim was to involve municipalities in the construction of a tourist network linked to the economy and culture of wine. In the following years, the creation of the *Movimento Turismo del Vino* (literally "Wine Tourism Movement") in 1993 and of the *Osservatorio Nazionale del Turismo del Vino* (literally "National Observatory of Wine Tourism") in 1999 gave a decisive boost to the phenomenon of wine tourism in Italy. They represented the beginning of a wide-ranging reflection about the best strategies to be implemented to modernise the image of Italian wine (Cinelli Colombini 2007): the debate in particular concerned the young people's consumption—they were no longer attracted to wine as it was considered an outdated product. At the end of the twentieth century some Italian wine circles highlighted that new fast-food as McDonald's and breweries were more attractive than the traditional "osteria" (tavern), which was in a phase of progressive decadence and remediable disappearance: the wine sector needed other sales locations and another youthful image to recover young customers. At the same time, the new links between quality wines, environmental promotion and tourism practices allowed to enlarge the market (Weaver 2001).

The *Movimento del turismo del vino* in particular organised the "Open Wine Cellars Days", which were well received by the public, demonstrating that the wine sector was able to intercept a demand for leisure and escape (Inglis 2000) from post-industrial society. In the first edition, which only covered Tuscany, the wineries involved were visited by about 150,000 people. In the last national editions before the Pandemic Covid-19, participants exceeded one million and proved the success of tourism events linked to wine (Cinelli Colombini 2015). The age of visitors to wine cellars dropped below 30 years old and news about "open wine cellar days" were the subject of mass media attention (Cinelli Colombini 2017): so, wine, a popular alcoholic beverage that was the cause of excesses and behaviour to be recriminated, acquired another dimension, less negative and much more suggestive, capable of stimulating a new demand among young people and women, who were both tourists and consumers.

Regardless of the results of the historiographical debate (Salvado 2016; Ben Tahar et al. 2021), the *Movimento del turismo del vino* was a happy intuition: it interpreted and channelled the growing demand for eco-friendly territories, and it experimented using a direct relationship with winemakers the link between wine production, history and culture. The rediscovery of wine in particular and of the rural world in general was the tourists' attempt to find new concerns and places of "refuge" for their leisure time: new values were given to the typicality and the sustainable environment. There was also a contradiction: the alcohol created a new market based on the demand for healthy products and the protection of ecosystems (Montella 2017).

In the new millennium a special attention was again dedicated to the routes of wine (but also oil, pasta, and, in general, flavours) because they were networks spread all over the Italian peninsula by involving a large number of small municipalities. The routes of flavours represented a tool allowing an evolution oriented towards both the upgrading and the development of the Italian territory. The wine route in particular contributed to the development of other sectors of the food cluster: tourism, agro-tourism, and food industry. This process promoted the development of less advanced areas, such as mountain areas experiencing a loss of population as well as of economic activities. In these areas there were most of the typical products (3200 local productions come from the Italian mountain areas): they influenced the social and economic development of local rural territories, in particular through the achievement of the following socio-economic benefits—the increase in farms income, a greater social vivacity, the valorisation and conservation of traditional activities, and the development of a food and wine tourism (Carlsen and Charters 2006). So, the routes of typical products included a very articulate set of initiatives, which involved public and private actors variously participating in a network favouring the local social and economic development. The main components of these routes were the products (wine, oil, pasta, cheese, rice, etc.), the territory (countryside, burgs), the ecosystem (landscape and environment quality enhancement), and the key players (producers, tour operators, associations, institutions). Before the Pandemic Covid-19, there existed 150 Italian wine routes, involving more than 1450 municipalities, 400 DOC/PDO wines, and 3300 wineries. These latter generated a turnover of around 2.5 million euros in 2009. However, many roads appeared (and continue to) inadequate to improve tourist flows due to a lack of means and funding. Besides, the great number of wine roads and the related fragmentation of initiatives prevented a correct allocation of available resources. So, even if a large number of wine routes existed in theory, only a very small group of routes were actually operational. Besides, there were relevance differences between the ways used by regions to finance and create wine routes (Fig. 5.2): at the top was Veneto (16), followed by Tuscany (14), Emilia-Romagna (13). Other regions were between 10 and 12 (Lombardy, Calabria, Apulia, and Campania) or between 5 and 9 (Trentino-Alto Adige, Piedmont, Friuli Venezia Giulia,

Fig. 5.2 Wine routes in Italy. (Source: © authors)

Umbria, Lazio, Marche, Sicily, and Sardinia). Less than 5 were in Valle d'Aosta, Liguria, Molise, and Basilicata.

The Italian model of multiplying wine routes at the regional level also depended on the need to meet the economic expectations of a plurality of public and private operators considering wine and food tourism as a relevant factor for the local and national economic development: someone also wanted to create wine districts. In other European countries having a consolidated wine-growing tradition, the wine routes were less: in Spain 17, in France 15, and in Portugal 11. This disparity was related to a different approach: in Italy the concept of "wine routes" gradually evolved towards the wider concept of "taste routes" (Vaquero Piñeiro et al. 2019)—this in fact allowed the inclusion of other typical products such as oil, apples, cheese, rice, salami, and so on. So, a dense network of food and wine routes was formed covering all the Italian countryside: going beyond wine, it was also possible to generate new expectations, soliciting the interests of a much more heterogeneous public.

Another aspect fuelling the relationship between viticulture and tourism (Sigala 2019) was the organisation of fairs and exhibitions presenting Italian wines: they progressively became real spin-offs in terms of tourism and advertising. In 1933, the first "Mostra Mercato dei Vini Tipici d'Italia" (literally "Exhibition Market of Typical Italian Wines") was organised in Siena: for the first time many of the best and traditional wines were presented together to the public. Siena was chosen because its province included a large part of the Chianti area: besides, it had the typically medieval atmosphere greatly appreciated by the Fascist regime. A huge crowd of visitors filled the city, nobody expected so many: the slogan created by the futurist poet Filippo Tommaso Marinetti was "Brunello is petrol!", that is, the Italian wine was the energy moving the world. As the Fascist regime use exhibitions for propaganda and self-celebration (Carli 2021), the space was dominated by very tall pavilions, scenarios studied down to the smallest detail: everything was modern, futuristic, and ambitious.

The wine exhibitions in Siena continued regularly after the Second World War: the city was in fact universally considered a perfect combination of Italy's historical past, the Middle Ages, and the Renaissance, and the world of wine. In 1950, the "Ente Autonomo per la Mostra Mercato Nazionale dei Vini Tipici" (literally "Autonomous Institut for the National Market Exhibition of Typical Wines") was founded and in 1960 the "Enoteca Italiana" was created with its headquarters in the imposing

Medici fortress in Siena: the aim of this public institution (abolished in 2017) was to spread the culture of Italian wine through exhibitions, tastings, and many other events favouring the wine tourism (Hall and Sharples 2008).

In 1967 Vinitaly, the international trade fair progressively becoming one of the most world important meetings of the wine sector, began in Verona (Capitello et al. 2014). Vinitaly provided an opportunity for specialists and producers to have information about trends in world markets, but it also attracted the interest of hundreds of people desiring to participate in wine and food tourism (Charters and Mitchell 2014). In the last editions, there were more than 200,000 visitors, demonstrating the economic and media force of wine and food-related events in the new millennium: these latter merged together countryside and city, family kitchen and starred cuisine, large markets and small shops, private companies and public policies. The *Fabbrica Italiana Contadina* (FICO, literally "Italian Peasant Factories"), the large thematic area inaugurated in Bologna, had one million visitors and the turnover was 50 million euros.

The evolution of wine tourism was strictly linked to the introduction of radical changes in winemakers' image and communication style: in a globalise and strongly competitive market it was in fact necessary to attract the attention of an increasingly demanding and refined international demand (Falabrino 2007). It was not enough to produce a good or excellent quality wine; it was necessary to have a good web reputation in order to transmit sensations (memories, images, thoughts) that obviously also came from the suggestions offered to customers by cinema, theatre, literature, and even art. The Italian wine sector had to carry out an incisive renewal operation presenting itself on the market by offering a wider range of products, suited to an increasingly young and cultured clientele. Initiatives such as open cellars, wine routes, and major wine events became tourist attractions and allowed both national and foreign tourists to discover Italian terroirs and enjoy a different experience, which went beyond simple tasting because it offered a direct contact with the rural world. For example, the Umbria was included in 2014 among the 10 Best Wine Travel Destinations by the US magazine *Wine Enthusiast*; besides, the ranking of tourist destinations proposed by the readers of *USA Today* inserted Tuscany among the Top 10 wine destinations and illustrated the four things that wine tourists most appreciated: natural environment, culture, gastronomy, and obviously high-quality wines. Tourists were also attracted by the myth of world-class wine landscapes and the presence of well-kept tourist services and of enveloping cultural atmospheres (Pezzi et al. 2021).

Faced with an evocative scenario, in which the agriculture filled itself with meanings and values going beyond the production of essential goods, wine companies' brands were involved in the new relationship between agriculture, the environment, and tourism (Smith 2006). The image of the Italian countryside was transformed: the long, parallel rows of vines characterising the viticulture became the perfect, almost paradigmatic expression of harmonious rural landscapes. Vineyards, perched on gentle hillsides and equated with pieces of furniture, were entrusted with the task of symbolising entire territories, of being the visual strength of a selected tourist offer (Image 5.2).

There were no more the old winemakers producing very "genuine" but usually not very "drinkable" wine: the Italian wine industry was represented by young entrepreneurship, with a relevant female presence (about

Image 5.2 Vineyards in Montefalco (Province of Perugia). (Source: Archivio Antonelli San Marco Società Agricola, Località San Marco, Montefalco (Italy))

30 per cent of wineries) (Matasar 2005), committed to offering a wide range of goods and services: from the direct sale of products to hotel accommodation, from the organisation of cultural events to the creation of a precise company profile. Wine cellars were no more used exclusively to store wine: they in fact had another function. Their architectural form had to stimulate tourists' interest and curiosity (Chiorino 2011) and so, most relevant wineries involved world-famous architects to create new buildings that combined productive efficiency and attraction for consumers and tourists. As the best winemakers in the early twentieth century represented in advertising messages their establishments as a symbol of modernity and progress, in the new millennium the best wine companies invited customers and specialised press to appreciate the modern and sometimes surprising design of their new wine cellars that symbolised high quality of the products combined with other luxury sensations and attention to the sustainability (Rocchi and Gabbai 2013). The "Petra" cellars by Mario Botta and "Rocca Frassinello" by Renzo Piano were concrete proof of this change in the use of wine cellars. The solutions adopted, as also shown by the Antinori group's wineries, in fact aimed to make the architectural intervention compatible with the protection and enhancement of the terroirs. It was evident that the involvement of the much-vaunted "archistars" simply represented an advertising operation that contributed nothing to the improvement of the wine quality: however, the hedonistic dimension achieved by the wine tourism assumed the same relevance of the high-quality wine. This last became only one aspect, and perhaps not the relevant one, for customers loving wine-growing landscapes and slow travel between food and wine, elegance and refinement of their settings.

Before Pandemic Covid-19, the wine tourism in Italy was growing at an annual rate of 12 per cent: in 2018, there were almost 14 million "wine tourism presences", a definition including a plurality of profiles and motivations (Charters and Ali-Knight 2002) which generated an economic value of between 2.5 and 3.5 billion euros. A relevant example of transformations connected to this tourism was represented the small Tuscan town of Montalcino taking advantage by "Brunello" image: between 2011 and 2016, tourist arrivals related to wine increased from 5000 to 24,000 and favoured a multiplication of hotels (from 14 to 50) and bed and breakfast facilities (from 78 to 620). Other Italian terroirs experienced a radical change in their image were Conegliano (Veneto), Oltrepò Pavese (Lombardy), the Langhe (Piedmont), and Montefalco (Umbria). The tourism model moved from a mass "sun and beach" tourist offer towards a more diversified idea of how to enjoy the free time which was strongly related to the attractiveness of the countryside and the consumption of typical products. This pushed more winemakers to adapt their structures transforming their simple places of production into suitable structures for tourist accommodation.

Italian visitors accounted for 62 per cent of the total number, while in terms of spending capacity, foreign tourists predominated, spending between 50 and 100 euros per capita, using visits to wineries to purchase quality products. Besides, they combined their wine tourism holiday with visits to the nearby areas and this obviously stimulated an integrated tourist including cultural, sporting and relaxation proposals,

Furthermore, in Italy there were 172,688 restaurants of which 586 of excellence, 11,632 farms with restaurants, 18,632 farms with accommodation, 170 wine routes and flavours, and 99 museums of taste (Reports on gastronomic tourism, data from 2016–2017). There were also more than 1200 wine cellars equipped to receive tourists: an extraordinary growth in comparison with the early 1990s, when few Italian winemakers thought that wine cellars could become a tourist destination in Italy, there were only 25 wine cellars open to the public with a total turnover of 200,000 euros. During individual visits and wine tours, they offered tastings, visits to the vineyards and other recreational and educational activities. Turnover linked to tourists visiting wine cellars increased by about a 31.35 per cent: it involved hotels, restaurants, and direct sales in the wineries. These percentages recalled those of the World Food Travel Association, according to which approximately 25 per cent of visitor spending were attributed to food and drink while travelling. This change took place in extra-European countries such as Chile, Australia, or New Zealand, where the wine tourism sector grew together with viniculture, while in Italy and in the European countries the transformation of wineries into tourist destinations has induced a business change in order to adapt buildings and rural settings to visiting activities (González 2019; Aravena Alvarado 2017; Medina-Albaladejo and Martínez-Carrión 2010; Carrau 1997). It was also necessary to offer the image of a sustainable production, attentive to the preservation of nature and to the application of the most modern technologies for recycling and saving energy.

More than 37 per cent of the wineries belonging to the "Wine Tourism Movement" offered overnight accommodation through a range of possibilities (farmhouses, mini-apartments, luxury relais, hotels in historic homes). Other accommodations were often provided by agri-tourisms and Bed and Breakfast facilities. Visits to wine cellars and wine tasting, combined with excursions, cookery courses and beauty treatments in fact created a varied tourist package. It is difficult to evaluate the real impact on wine sales because it is necessary to add the sales during the visits to those arriving after when satisfied and happy tourists come back home and talked to their friends about their wine tourism experiences.

This new wine tourism also had a relevant characteristic meeting the favour of European institutions: it did not concentrate visits to wine cellars in a single period of the year, that is it covered the whole year. European institutions in fact had a negative evaluation of the traditional concentration of the tourist flows in a limited period of the year because this favoured the congestion of transport and reception services and quality of the environment. So, the EU fifth framework programme for the environment and sustainable development established a strategy including a policy of "deseasonalisation" of tourist movements.

Before Pandemic Covid-19, data concerning wine tourism indicated that the month preferred by wine tourists was May, followed by August, September, July, and June. In the winter wineries were mainly visited by people living in nearby towns and villages, autumn was favoured by day trippers, while spring was characterised by day visits and tourists staying overnight. Winery was not just a weekend destination: visits were spread over weekdays as well. Wine tasting was obviously the most popular activity, combined with cooking classes and guided tours of the vineyards and of the wineries, in particular in the case of architectural complexes of particular historical value. Conferences and the tasting of local products represented a relevant attraction for more than 50 per cent of tourists. Finally, the modern wine tourist was profiled: she/he was between 30 and 50 years of age (so a widest range), with a high cultural level, an Internet user, and inclined to do-it-yourself trips without relying on travel agencies. This obviously means that wine companies and agencies operating in wine tourism had to dedicate an attentive care to the evolution of her/his behaviours and tastes.

It was clear to observers at the end of the twentieth century that industrial development, from the point of view of food, had generated certain phenomena such as the loosening or elimination of economic and cultural ties between food and territory, the uniformity of food models and their new urban characterisation. It was argued that the cultural dimension inherent in food was in danger of disappearing, buried under the impositions of the mass market. The connection between agri-food products and reference territory was destined to be broken forever (Pastore 2016). However, compared to a general panorama dominated by some "heavy trends" heavily aimed at homologation, a "soft trend" was beginning to be glimpsed as a counter-trend. This was a change of course that was still very weak and tended to give new meaning to food, its territorial provenance and its historical-cultural and environmental location.

As regards the economic impact of climate change on viticulture, the most recent studies show that climate change will have both positive and negative effects on the various terroirs. In the short term, changes in particular will concern productive yields and wine quality: this implies that prices and wine companies returns will change too (Jones et al. 2005). In the long term, strong variations concerning resources for the wine production are expected, especially in the Mediterranean wine regions: significant changes in the "geography" of wine production will be possible and will surely involve the Italian peninsula. EU legislative interventions and winemakers' productive choices aim to reduce or balance out the worst consequences of climate change (however, the final results have a low relation with European pollution and strongly depend on other countries' political decisions, moreover China and the United States). For the wine sector the most promising solutions are coming from the biodiversity and the diversification of the varieties of grapes. Climate change will permit grape growing at high altitudes in mountainous regions (Mozell and Thach 2004). Therefore, what is today called "heroic viticulture" may soon be a normal and widespread winemaking practice in the valleys of the Alps and the Apennines. On the contrary, relevant problems related to the greatest climate changes will concern the wines in the Italian Southern regions (in particular Sicily): the excessive alcoholic degree and the arrival of new parasites will strongly modify wine production and oblige winemakers to sell their products as blending wines or transform them in spirits and liquors (Image 5.3).

Image 5.3 Grape picker with a pack basket full of grapes in the Valtellina valley (1960s). (Source: Archivio Storico Unione Italiana Vini—Milano)

5.4 Italian Winemaking and Covid: First Tentative of Reflections

The global health crisis generated by Pandemic Covid-19 obviously had negative effects on the Italian wine sector provoking a general contraction of sales: the reduction in domestic consumption and the difficulties in exporting were worsened by the almost total paralysis of tourism. So, 60 per cent of companies expected a drop in export sales by 2020 and, within these, 37.5 per cent expected the drop to be more than 10 per cent. This is a worse picture than that of 2009 due to the 2008 crisis, when 60.6 per cent of wine businesses suffered a drop in sales, with a drop in turnover of 3.7 per cent and falls of over 10 per cent affecting 24.2 per cent of businesses.

However, the Italian wine sector was able to implement new marketing strategies and partially reduced the overall impact of the Pandemic Covid-19.

Exporters and Importers of Wines, Spirits, Liqueurs, Syrups, Vinegars, and Similar Products (FEDERVINI) in March 2021 highlighted that wine exports reached 6.29 billion euros in 2020, a result that marks a -2.2 per cent compared to 2019, but a +0.8 per cent compared to 2018. This outcome was the result of contrasting trends between the various periods of the year: 2020 had in fact started extremely positively compared to the already excellent 2019, with a +5.2 per cent in the first quarter (€1.51 billion in the first three months of the year compared to €1.44 in 2019), while from the following month the effects of the Covid-19 emergency began to be felt more strongly with a sharp drop in exports (in May it reached -24.3 per cent) for an overall result of -12.6 per cent in the second quarter (€1.38 billion in 2020 compared to €1.57 billion in 2019). The third quarter with -1.9 per cent showed the beginning of a "resilience period", with an overall result of 1.56 billion euros compared to 1.59 billion euros in 2019. Some positive signs arrived in the fourth quarter of 2020 with +0.7 per cent, on the strength of exports that grew strongly in November and December: the overall result of over 1.84 billion euros compared to 1.83 billion euros in 2019.

The difficulties were in particular in Asian markets (-12.6 per cent) and American one (-4.3 per cent), while exports to European countries held up (+0.2 per cent) which together account for 62 per cent of Italian wine exports worldwide. Among the top five Italian regions for wine exports, three showed a growing export result in 2020: Piedmont (+2.6 per cent), Trentino-Alto Adige (+4.3 per cent), and Emilia-Romagna (+3.4 per cent). In opposite, Veneto (-3.3 per cent) and Tuscany (-3.2 per cent) were in moderate decline, while the regions with the biggest drop in wine exports were Sicily (-11.9 per cent), Lombardy (-11.7 per cent), Friuli-Venezia Giulia (-8.9 per cent), and Apulia (-7.6 per cent). Concerning online sales, a relevant increase was registered: +110.2 per cent. A great growth also concerned young consumers between 18 and 24 years old (+209.2 per cent) and the over 65s (+27 per cent). Men's purchases grew up 100.6 per cent, while women's one were 65.3 per cent higher.

Concerning the winemakers, important entrepreneurial strengthening operations continued. The Tuscan Antinori group took control of Jermann, a renowned wine producer in Friuli. The Antinori-Jermann joint venture became a twinning of a family brand that closed its balance sheet with 178 million euros in 2020 and a company, also family-owned, with around 15 million euros in revenues, arriving thanks to a relevant presence in Italian and all main world markets (in particular in German and US ones). Besides, in early 2021, the merger of the Custoza and Valpantena cooperative wineries created the Cantine di Verona, a consortium with a turnover of 60 million euros and a production of 30 tonnes of grapes.

Wine associations also called for immediate action from the institutions, together with longer-term measures. Many experts, including Ernesto Abbona, president of the "Unione Italiana Vini", asked for a reform of the Common Agricultural Policy including the national support plan for the wine sector to avoid structural crisis and to relaunch the Italian exhibition system—Vinitaly in primis—through the "Pact for exports". To avoid that the pandemic crisis could become a structural crisis for the Italian sector, they required other interventions: the digital objective, the strengthening of tools to ensure certainty of production and declared data, the reform of the National Wine Committee and of the DOC/IGT system and its control structure, the new dialogue with the United States to put an end to the season of duties and tensions on trade policies, the relaunch of tourism with economic aids for tourists, the improvement of infrastructures favouring the distribution of rural products, wines included. In this regard, they also observed that in Italy the EU Next Generation provided resources of only 8 billion out of a total of 210 billion and asked for more European funds for the ecological transition and sustainable production involving the viticulture. Furthermore, they highlighted the relevance for Italian vineyard and wines of the strengthening of broadband in all agricultural territories (5G) for its obvious links with the wine e-commerce and tourism.

References

Albala, Ken, ed. 2013. *Routledge International Handbook of Food Studies*. London and New York: Routledge.

Anderson, Kym, and Vicente Pinilla, eds. 2018. *Wine Globalization. A New Comparative History*. London and New York: Cambridge University Press.

Antonioli Corigliano, Magda. 1999. The Route to Quality: Italian Gastronomy Networks in Operation. In *Tourism and Gastronomy*, ed. Greg Richards and Anne-Mette Hjalager, 166–185. New York and London: Routledge.

Aravena Alvarado, Rodrigo, ed. 2017. *Chilean Wine-Growing Heritage: Approaches to Wine Culture*. Santiago de Chile: Ediciones Biblioteca Nacional.

Asero, Vincenzo, and Sebastiano Patti. 2009. *From Wine Production to Wine Tourism Experience: The Case of Italy*. American Association of Wine Economists, Working Paper, 52.

Balzani, Roberto, ed. 2015. *I territori del patrimonio: dinamiche della patrimonializzazione e culture locali (secoli XVII–XX)*. Bologna: il Mulino.

Battilani, Patrizia. 2001. *Vacanze di pochi, vacanze di tutti*. Bologna: il Mulino.

———. 2010. L'americanizzazione del turismo europeo. In *La rivoluzione dei consumi. Società di massa e benessere in Europa 1945–2000*, a cura di. S. Cavazza and E. Scarpellini, 221–254. Bologna: il Mulino.

Ben Tahar, Yosr, Coralie Haller, and Charlotte Massa. 2021. Business Tourism in the Wine Sector: An Exploratory Study. *Journal of Wine Research*. https://doi.org/10.1080/09571264.2021.2012442.

Benatti, Francesco Emanuale, and Paolo Castelletti, eds. 2022. *Dai vini tipici al testo unico. Profilo storico della normativa delle denominazioni di origine del vino in Italia (1904–2021)*. Milano: Unione Italiana Vini.

Berrino, Annunziata. 2011. *Storia del turismo in Italia*. Bologna: Il Mulino.

Boatto, Vasco, and Federica Bianchin. 2011. L'enoturismo e l'enoturista: lo stato dell'arte. In *La roadmap del turismo enologico*, ed. Vasco Boatto and Alejandro Juan Gennari, 13–23. Milano: FrancoAngeli.

Cantino, Valter, Massimo Pollifroni, Damiano Cortese, and Mariangela Longo. 2016. Branding Strategies and Accounting in Wine Production: A Historical Analysis of Fontanafredda (1858–1932). In *Accounting and Food. Some Italian Experiences*, ed. Massimo Sargiacomo, Luciano D'Amico, and Roberto Di Pietra, 23–41. New York: Routledge.

Capitello, Roberta, et al. 2014. Business-to-business Service Quality Assessment in Professional Wine Events: The Case of Vinitaly. In *Food and Wine Events in Europe. A Stakeholder Approach*, ed. Alessio Cavicchi and Cristina Santini, 110–124. London and New York: Routledge.

Carli, Maddalena. 2021. *Vedere il fascismo. Arte e politica nelle esposizioni del regime (1928–1942)*. Roma: Carocci.

Carlsen, Jack. 2004. A Review of Global Wine Tourism Research. *Journal of Wine Research* 15 (1): 5–13.

Carlsen, Jack, and Steve Charters, eds. 2006. *Global Wine Tourism: Research, Management and Marketing*. Cambridge: Cabi.

Carrau, Francisco M. 1997. The Emergence of a New Uruguayan Wine Industry. *Journal of Wine Research* 8: 179–185.

Cavaco Carminda, M.M. 1995. Rural Tourism: The Creation of New Tourist Spaces. In *European Tourism. Regions, Spaces and Restructuring*, ed. A. Montanari and A.M. Williams, 127–150. Chichester: Wiley.

Cavazza, Stefano. 1997. *Piccole patrie. Feste popolari tra regione e nazione durante il fascismo*. Bologna: Il Mulino.

Cavazza, Stefano, and Emanuela Scarpellini, eds. 2010. *La rivoluzione dei consumi. Società di massa e benessere in Europa 1945–2000*. Bologna: il Mulino.

Charters, Steve, and Jane Ali-Knight. 2002. Who is the Wine Tourist? *Tourism Management* 23 (3): 311–319.

Charters, Steve, and Richard Mitchell. 2014. Food and Wine Events in Europe and the New World. In *Food and Wine Events in Europe. A Stakeholder Approach*, ed. Alessio Cavicchi and Cristina Santini, 15–27. London and New York: Routledge.

Charters, Steve, and Nathalie Spielmann. 2014. Characteristics of Strong Territorial Brands: The Case of Champagne. *Journal of Business Research* 67 (7): 1461–1467.

Chiorino, Francesca. 2011. *Cantine secolo XXI. Architetture e paesaggi del vino*. Milano: Mondadori Electa.

Cinelli Colombini, Donatella. 2007. *Il marketing del vino e del territorio: istruzioni per l'uso*. Roma: Agra.

———. 2015. Wine Tourism in Italy. *International of Tourism Research* 7: 29–35.

———. 2017. *Marketing delle cantine aperte*. Roma: Agra.

Clancy, Michael, ed. 2018. *Slow Tourism, Food and Cities. Pace and the Search for the "Good Life"*. New York: Routledge.

Conz, Elisa, and Giovanna Magnani. 2019. Brand Identity of Long Lasting Family Firms in the Wine Industry. *Micro e Macro Marketing* 28 (1): 139–158.

Corsi, Alessandro, Simonetta Mazzarino, and Eugenio Pomarici. 2019. The Italian Wine Industry. In *The Palgrave Handbook of Wine Industry Economics*, ed. Adeline Alonso Ugaglia et al., 47–76. Cham: Palgrave Macmillan.

Costantino, S., and A. Artista. 2003. *Le strade del vino e le vie dello sviluppo*. Milano: FrancoAngeli.

Crescenzi, Riccardo, Fabrizio De Filippis, Mara Giua, and Cristina Vaquero Piñeiro. 2021. Geographical Indications and Local Development: The Strength of Territorial Embeddedness. *Regional Studies*. https://doi.org/10.1080/00343404.2021.1946499.

Da Silva Lopes, Teresa. 2007. *Global Brands: The Evolution of Multinationals in Alcoholic Beverages*. Cambridge: Cambridge University Press.

De Leo, Carolina, and Giovanni Favero. 2018. *Ca' Foscari e Carpenè Malvolti. Il Risorgimento dell'economia nel Veneto dell'Ottocento*. Venezia: Università Cà Foscari.

Decleva, Enrico, ed. 2001. *Ulrico Hoepli 1847–1935. Editore e libraio*. Milano: Ulrico Hoepli.

Falabrino, Gian Luigi. 2007. *Storia della pubblicità in Italia dal 1945 a oggi*. Roma: Carocci.

Fasce, Ferdinando. 2012. *Le anime del commercio. Pubblicità e consumi nel secolo americano*. Roma: Carocci.

Fauri, Francesca, and Paolo Tedeschi, eds. 2011. *Novel Outlooks on the Marshall Plan. American Aid and European Re-Industrialization*. Bruxelles: Peter Lang.

Fennell, David A. 1999. *Ecotourism. An Introduction*. New York: Routledge.

———. 2013. Ecotourism. In *The Routledge Handbook of Tourism and the Environment*, ed. Andrew Holden and David Fennell, 323–334. New York: Routledge.

Fennell, David A., and David Weaver. 2008. The Ecotourium Concept and Tourism-Conservation Symbiosis. *Journal of Sustainable Tourism* 13 (4): 373–390. https://doi.org/10.1080/09669580508668563.

Gabrielli, Patrizia. 2011. *Anni di novità e di grandi cose. Il boom economico fra tradizione e cambiamento*. Bologna: il Mulino.

Galletto, Luigi, and Federica Bianchin. 2011. L'enoturismo tra Vecchio e Nuovo Mondo. In *La roadmap del turismo enologico*, ed. Vasco Boatto and Alejandro Juan Gennari, 24–57. Milano: FrancoAngeli.

Gmelch, George, and Sharon Gmelch. 2011. *Tasting the Good Life: Wine Tourism in the Napa Valley*. Bloomington: Indiana University Press.

González, Pablo Alonso. 2019. La vuelta al terroir: el despertar de la cultura del vino en España. RIVAR (Revista Iberoamericana de Viticultura, Agroindustria y Ruralidad. Editada por el Instituto de Estudios Avanzados (IDEA) de la Universidad de Santiago), n. 6: 62–89.

Govoni, Paola. 2011. *Un pubblico per la scienza. La divulgazione scientifica nell'Italia en formazione*. Roma: Carocci.

Grazioli, Elio. 2001. *Arte e pubblicità*. Milano: Bruno Mondadori.

Guy, Kolleen M. 2003. *When Champagne. Became French: Wine and the Making of National Identity*. Baltimore: Johns Hopkins University Press.

Hall, Colin Michael, and Liz Sharples. 2008. *Food and Wine Festivals and Events around the World: Development, Management and Markets*. Oxford: Elsevier.

Hall, Colin Michael, Liz Sharples, Brock Cambourne, and Nik Macionis, eds. 2000. *Wine Tourism around the World. Development, Management and Markets*. New York: Routledge.

Harvey, Matt, Warwick Frost, and Leanne White. 2014. Exploring Wine and Identity. In *Wine and Identity. Branding, Heritage, Terroir*, ed. Matt Harvey, Warwick Frost, and Leanne White, 17–28. New York: Routledge.

Higgins, David M. 2018. *Brands, Geographical Origin and the Global Economy: A History from the Nineteenth Century to the Present*. Cambridge: Cambridge University Press.

Holden, Andrew, and David Fennell, eds. 2013. *The Routledge Handbook of Tourism and the Environment*. New York: Routledge.

Inglis, Fred. 2000. *The Delicious History of the Holiday*. New York: Routledge.

Inglis, David, and Anna-Mari Almila, eds. 2019. *The Globalization of Wine*. London: Bloomsbury USA Academic.

Jones, Gregory V., Michael A. White, Owen R. Cooper, and Karl Storchmann. 2005. Climate Change and Global Wine Quality. *Climatic Change* 73: 319–343.

Kernohan, David. 2014. Wine and Architecture-Structure and Elegance. In *Social, Cultural and Economic Impacts of Wine in New Zealand*, ed. Peter J. Howland, 137–152. New York: Routledge.

Lacoste, Pablo, and Jorge Zamora. 2007. Tourism and Wine: A Marriage of Convenience or True Love? *Journal of Wine Research* 18 (2): 121–123.

Lasansky, Medina. 2004. *The Renaissance Perfected: Architecture, Spectacle and Tourism in Fascist Italy*. Philadelphia: Penn State University Press.

Lazzaro, Claudia, and Roger Crum. 2005. *Donatello among the Blackshirts: History and Modernity in the Visual Culture of Fascist Italy*. Ithaca: Cornell University Press.

Magliacani, Michela. 2015. *Managing Cultural Heritage: Ecomuseums, Community Governance, Social Accountability*. Cham: Palgrave Macmillan.

Marescalchi, Arturo. 1904. *G.A. Ottavi e I 50 anni del "coltivatore"*. Casale Monferrato: tip. Carlo Cassone.

Matasar, Anna B. 2005. *Women of Wine. The Rise of Women in the Global Wine Industry*. Berkeley: University of California Press.

Mazza, Marta. 2010. Dalla Belle époque alla Grande guerra: frammenti di vita italiana in pubblicità. In *Manifesti. Pubblicità e vita italiana 1895–1945*, ed. di Anna Villani, 62–85. Cinisello Balsamo, Milano: Silvana Editoriale.

Medina-Albaladejo, Francisco J., and José M. Martínez-Carrión. 2010. Change and Development in the Spanish Wine Sector, 1950–2009. *Journal of Wine Research* 21: 77–95.

Montanari, Armando. 1995. The Mediterranean Region: Europe's Summer Leisure Space. In *European Tourism. Regions Spaces and Restructuring*, ed. Armando Montanari and Allan E. Williams, 41–66. Chichester and New York: Wiley.

Montella, Marta Maria. 2017. Wine Tourism and Sustainability: A Review. *Sustainability* 9 (1): 113. https://doi.org/10.3390/su9010113.

Mora, Pierre. 2016. *Wine Positioning: A Handbook with 30 Case Studies of Wine Brands and Wine Brands and Wine Regions in the World*. Cham: Springer.

Mozell, Michelle Renée, and Liz Thach. 2004. The Impact of Climate Change on the Global Wine Industry: Challenges & Solutions. *Wine Economics and Policy* 3: 81–89.

Navarrini, Gianmarco. 2015. *I mondi del vino. Enografia dentro e fuori il bicchiere*. Bologna: il Mulino.

Pastore, Riccardo. 2016. *Il marketing del vino e del territorio: istruzioni per l'uso*. Milano: FrancoAngeli.

Pezzi, Maria Giulia, Alessandra Faggian, and Neil Reid. 2021. *Agritourism, Wine Tourism, and Craft Beer Tourism: Local Responses to Peripherality through Tourism Niches*. New York: Routledge.

Pinilla, Vicente, and María Isabel Ayuda. 2007. ¿Un asunto mediterráneo? El mercado internacional del vino, 1850–1936. *Territorios del vino* 1 (1): 8–26.

Resnick, Evelyne. 2008. *Wine Brands: Success Strategies for New Markets, New Consumers and New Trends*. New York: Palgrave Macmillan.

Richards, Greg. 2002. Gastronomy: An Essential Ingredient in Tourism Production and Consumption. In *Tourism and Gastronomy*, ed. Greg Richards and Anne-Mette Hjalager, 3–20. New York and London: Routledge.

———. 2015. Evolving Gastronomic Experiences: From Food to Foodies to Foodscapes. *Journal of Gastronomy and Tourism* 1: 1–14.

Rocchi, Benedetto, and Manuela Gabbai. 2013. Territorial Identity as a Competitive Advantage in Wine Marketing: A Case Study. *Journal of Wine Research* 24 (4): 291–310.

Salvado, Josefina. 2016. Wine Culture, Territory/Landscape and Tourism, the Enotourism Key Pillars. How to Get Business Success and Territorial Sustainability Inside Wine Tourism Ecosystem? http://193.136.79.105:8080/bitstream/11328/1628/1/PROCEEDINGS per cent20BOOKS per cent20OF per cent20THE per cent20EATSA per cent20CONFERENCE per cent202016-per cent20josefina.pdf.

Scarpellini, Emanuela. 2011. *Material Nation: A Consumer's History of Modern Italy*. New York: Oxford University Press.

Schamel, Günter. 2009. Dynamic Analysis of Brand and Regional Reputation: The Case of Wine. *Journal of Wine Economics* 4 (1): 62–80.

Sigala, Marianna. 2019. Developing and Branding a Wine Destination Through UNESCO World Heritage Listing: The Case of the Mount Lofty Ranges Agrarian Landscape. In *Wine Tourism Destination Management and Marketing. Theory and Cases*, ed. Marianna Sigala and Richard N.S. Robinson, 99–112. Cham: Palgrave Macmillan.

Sigala, Marianna, and Richard N.S. Robinson, eds. 2019. *Wine Tourism Destination Management and Marketing. Theory and Cases*. Cham: Palgrave Macmillan.

Smith, Willian L. 2006. Experiential Tourism around the World and at Home: Definitions and Standards. *Journal of Services and Standards* 2 (1): 1–14.

Storchmann, Karl. 2018. Wine Economics. In *Handbook of the Economics of Wine Volume 1: Prices, Finance, and Expert Opinion*, ed. Orley Ashenfelter, Olivier Gergaud, Karl Storchmann, and William Ziemba, 3–40. Singapore: World Scientific Books.

Tedeschi, Paolo. 2011. La Politique Agraire Commune ou "le grand cartel agricole communautaire". In *Regulierte Märkte: Zünfte und Kartelle / Marchés régulés: corporations et cartels*, ed. M. Müller, H.R. Schmidt, and L. Tissot, 243–259. Zurich: Cronos Verlag.

Tiefenbacher, John P. 2013. Themes of U.S. Wine Advertising and the Use of Geography and Place to Market Wine. *EchoGéo* 23. http://echogeo.revues.org/13378; https://doi.org/10.4000/echogeo.13378.

Tiefenbacher, John P., and Christi Townsend. 2020. The Semiofoodscape of Wine: The Changing Global Landscape of Wine Culture and the Language of Making, Selling, and Drinking Wine. In *Handbook of the Changing World Language Map*, ed. Stanley D. Brunn and Roland Kehrein, 4103–4145. Springer.
Vaquero Piñeiro, Manuel, and Luciano Maffi. 2018. The Diffusion of Italian Wine in the United States (1861–1914). *Rivar. Revista Iberoamericana de viticultura, agroindustria e ruralidad* 15: 176–196.
Vaquero Piñeiro, Manuel, Paola de Salvo, and Francesca Giommi. 2019. *Rural Tourism and Territorial Development in Italy*. In *Sustainable Rural Development*, a cura di Florin-Costantin Mihai et al. London: IntechOpen.
Visentin, Francesco, and Francesco Vallerani. 2018. A Countryside to Sip: Venice Inland and the Prosecco's Uneasy Relationship with Wine Tourism and Rural Exploitation. *Sustainability* 10 (7). https://doi.org/10.3390/su10072195.
Walker, Larry. 2005. *The Wines of the Napa Valley*. London: Hachette.
Weaver, David Bruce. 2001. Introduction to Ecotourism. In *The Encyclopedia of Ecotourism*, ed. David Bruce Weaver, 1–3. Wallingford, UK: CABI Publishing.
Yeung, Peter, and Liz Thach. 2019. *Luxury Wine Marketing: The Art and Science of Luxury Wine Branding*. Oxford: Infinite Ideas.

Index

M. Vaquero Piñeiro et al., *A History of Italian Wine*,
https://doi.org/10.1007/978-3-031-06097-7

Z

The manufacturer's authorised representative in the EU is Springer Nature Customer Service Centre GmbH, Europaplatz 3, 69115 Heidelberg, Germany. If you have any concerns regarding our products, please contact ProductSafety@springernature.com

Printed and bound by CPI Group (UK) Ltd, Croydon, CR0 4YY
30/09/2026
02225321-0006